D1131167

Applications of Plasma Source Mass Spectrometry

Applications of Plasma Source Mass Spectrometry

Edited by
Grenville Holland
Department of Geological Sciences, University of Durham
Andrew N. Eaton
VG Elemental Ltd., Winsford

Selected papers from the Second International Conference on Plasma Source Mass Spectrometry held at the University of Durham, 24–28th September 1990 and sponsored by VG Elemental Ltd.

ISBN 0-85186-566-6

A catalogue record for this book is available from the British Library

Published by The Royal Society of Chemistry,
Thomas Graham House, Science Park, Cambridge
CB4 4WF

Printed in Great Britain by Bookcraft (Bath) Ltd.

Preface

Although mass spectrometry has long been established, methods involving plasma source mass spectrometry are still in their infancy. Even so, the techniques of inductively coupled plasma, and glow discharge, mass spectrometry have already revolutionised elemental analysis, with their advances in detection power, speed and simplicity of use. Analysts in many different scientific disciplines have been swift to recognise the importance of these new techniques and have readily adopted, and adapted, them to their particular needs, often in novel and innovative ways.

The common roots of these techniques led to the establishment in 1988 of a forum, the First Durham International Conference on Plasma Source Mass Spectrometry. At this forum scientists were able to meet and exchange ideas from a broad range of disciplines and share their knowledge and experience of plasma source mass spectrometry. Although a handful of papers from that first conference were eventually published no unified record of its proceedings was produced. As organisers we were determined that the many pertinent and interesting papers and posters presented at the Second Durham International Conference would survive and be published. This book, therefore, serves as a record of the proceedings and as a contribution to the literature on these new and rapidly evolving techniques.

Although the papers have all been independently refereed, as editors it has been our intention wherever possible to allow authors the opportunity to preserve their ideas and observations intact. The papers contained in this book, therefore, do not necessarily coincide with our own prejudices nor with those of the referees; for such prejudices can, with remarkable frequency, destroy the germ of a thought-provoking hypothesis or constrain the development of a new line of thought. The papers represent, with as much fidelity as possible, the views of the authors and their associates.

We would like to thank all those who contributed to the Second Durham International Conference and who subsequently submitted papers for publication in this book. We also extend our personal thanks to Dr. Robert Hutton and Miss Karen Morton at VG Elemental for their boundless support, help and advice. Finally we wish to thank Miss Catherine Lyall of the Royal Society of Chemistry for her patience and forbearing.

Grenville Holland
University of Durham

Andrew N. Eaton
VG Elemental

Contents

Glow Discharge Mass Spectrometry - Aspects of a Versatile Analytical Tool

D. Stuewer

INSTITUT FÜR SPEKTROCHEMIE UND ANGEWANDTE SPEKTROSKOPIE, POSTFACH 10 13 52, D-W4600 DORTMUND 1, GERMANY

1 INTRODUCTION

In the recent development of instrumental techniques for chemical analysis, mass spectrometry (MS) with plasma sources has gained a more and more dominant role. The inductively coupled plasma (ICP), well known as an excitation source in atomic emission spectrometry, has been established as an effective ion source for MS, and ICP-MS, mainly applied for solution analysis, has found wide-spread acceptance in the analytical community. For the sputtered neutrals mass spectrometry (SNMS), mainly applied for surface and in-depth analysis, acceptance is somewhat hesitant though it shows high depth resolution and convincing quantitative results. Both these techniques have been almost exclusively realized in the low resolution version with a quadrupole as mass analyzer. In case of glow discharge mass spectrometry (GDMS), the development was different, as the high resolution technique was realized first in a commercial system, very recently followed by commercial low resolution systems. The domain of GDMS is the direct elemental analysis of inorganic solids restricted to conducting and semiconducting solids. For reason of completeness, the application of laser mass spectrometry to local and micro distribution analysis should also be mentioned here, for which considerable progress is expected in the near future.

GDMS combines the basic advantages of MS techniques in a very special way. It enables a true multielement procedure which is executed quasisimultaneously. Low and nearly equal detection limits can be achieved for a great variety of elements. Data acquisition and processing is easy and can make full use of the information encoded in the spectra. The isotope dilution technique may be applied for quantitative measurements. Sample preparation is very straightforward. Finally it is a very versatile method which may be applied to microchemical analysis as well as to trace analysis and to surface and in-depth analysis as well as to bulk analysis.

2 BASIC ASPECTS

The basic task of an MS ion source is to atomize a solid and to subsequently ionize the atoms. If we succeed in realizing these two processes in individual and independent steps, the ionization efficiency or the final analytical sensitivity is determined by the simple product of both probabilities, and we will not suffer from non-spectral interferences such as matrix effects. The glow discharge realizes a good, but nevertheless incomplete approximation of this ideal situation.

The general voltage-to-current characteristic of a dc gas discharge shows two plateaus, in which the voltage remains nearly constant with increasing current. The first plateau corresponds to the Townsend discharge and the second to the normal glow discharge. This is characterized by the fact that the burning spot of the discharge does not cover the whole surface area of the cathode. An increase in the current results in an increase of the burning spot area while the voltage remains nearly constant. When the whole cathode surface is covered, the discharge changes from the normal state to the anomalous state in which a further increase of the current results in an increase of the voltage up to the final voltage break-through.

The glow discharge is a self-sustaining, space charge dominated discharge taking place in a working gas of reduced pressure. Concerning the structure of the discharge, it is only important here that we have a voltage drop region above the cathode, the cathode fall, followed by the negative glow beginning with the glow edge. A detailed consideration may find up to nine light emitting light regions and dark spaces in turn. The position of the anode is not significant as far as it is outside the cathode fall region. Inside, we have the hindered or obstructed discharge, for which the voltage must be increased further to avoid the discharge being extinguished.

In the cathode fall region, positive ions are attracted by the electrical field towards the cathode where they impinge on the surface generating an impact cascade. If a surface atom gets enough energy to leave the solid ensemble, it is sputtered and takes part in the discharge processes. If it is ionized in the voltage drop region - for instance by electron impact ionization - it will return to the surface. Beyond the cathode fall, we have a quasineutral plasma forming a reservoir for the extraction of ions. Beside electron impact ionization, the most important process is Penning ionization. In this process, a metastable of the working gas may transfer its excitation energy radiationless to a matrix atom which can be ionized if only its ionization energy is lower than the excitation energy of the metastable. In conclusion, analyte atoms sputtered from the surface of the sample can be extracted into the mass spectrometer if they are ionized in the glow region of the discharge. This is of course an oversimplification, but within the validity of this approximation, it demonstrates that atomization by sputtering from the surface and subsequent ionization in the region beyond the glow edge are independent, so that we can expect GDMS to be nearly free of matrix effects.

There are a lot of processes and mechanisms contributing to what is going on in the discharge, and charge transfer reactions also play an important role. The present state of our knowledge about the contribution of different mechanisms is really not comprehensive - neither experimentally nor theoretically.

An interesting paper concerning these problems has recently been published by Hess et al.[1] who used laser irradiation to depopulate metastable levels of the working gas. The response signal was measured by opto-galvanic spectrometry and by MS. The opto-galvanic signal shows an ionization increase due to enhancement of electron impact ionization in the very first moment followed by a negative signal due to the depopulation of metastable levels which is also represented in the negative peaks resulting as difference of the two spectra without and with laser irradiation. This technique enables a detailed study of the role of metastables in ion production.

3 EQUIPMENT

Figure 1 shows a sketch of a laboratory GDMS system which has been in use for several years[2]. It is a low resolution system using an obstructed discharge corresponding to the well-known Grimm type lamp. Ion extraction is carried out cathodically. The sample can be applied as a disc, pin or hollow cathode.

The ion source is completed by an exit aperture. This aperture should be very near to the glow edge in order to sample ions from the hot inner region of the plasma, avoiding the appearance of molecular ions from cooler zones influenced by wall effects. In the source, we have a reduced pressure of several hundred Pa. Pressure reduction to the high vacuum in the analyzer is achieved by a two stage differential pumping system. The following energy analysis is done here by a tubular lens system corresponding to a Bessel-box filter. For high resolution, a spherical condenser is applied. For mass analysis, low resolution systems use a quadrupole filter with unit resolution so that mass differences of at least 1 Da can be resolved. For the high resolution system, a magnetic field device serves as mass analyzer. Faraday cups and electron multipliers of different type are used for ion detection in analog or counting mode.

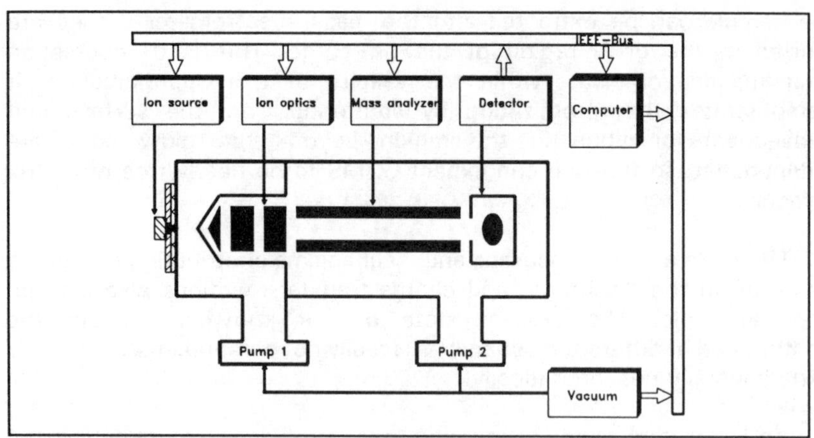

Fig. 1: Sketch of GDMS system with quadrupole

The first commercial system in the market was the VG 9000 (VG Instruments, Manchester, UK) which appeared in 1983 as a high resolution system. In 1989 the VG GloQuad followed as the first commercial low resolution system. Further systems have been announced by Extrel, Pittsburgh, USA, and Turner Scientific, Appleton, UK. For the latter, it should be emphasized here that it realizes the combination of GDMS and ICP-MS in one system.

4 BULK ANALYSIS

GDMS is well established for routine bulk analysis of pure and ultrapure materials as they are used in the semiconductor industry. Mykytiuk et al. have recently published a survey of the performance of GDMS in this field[3]. GDMS has become the method of choice for the routine analysis of ultrapure Al as it is used for sputter targets. The main reason is that no other technique combines true multielement facilities with the extremely low detection limits required in particular for U and Th. A comprehensive survey of Al analysis by means of GDMS has been published by Vassamillet[4] and by Kudermann[5]. Routine analysis of very pure Al comprises 30 elements; further 40 elements are determined as occasion demands.

The lowest detection limts for U and Th are achieved with neutron activation analysis (NAA)[5]. However, a substantial advantage of GDMS in comparison to NAA is that there is neither the need for a reactor nor any delay in getting the results. Of course there are problems in GDMS concerning precision and accuracy, for instance for the determination of Si due to its precipitation at grain boundaries and of gases for which the determined value is at best an upper limit of the true content indicating the state of cleanliness of the plasma cell.

The analytical performance of a low resolution system has been characterized by the analysis of steel standards[6], taking into account low alloy steel as well as high alloy steel. In a multielement procedure with detection limits of about 0.3 μg/g a relative precision between 0.01 and 0.05 has been achieved for the majority of 30 elements; in a test analysis, the mean relative deviation between determined values and values taken from the certificate was less than 0.1.

5 INTERFERENCES

The main obstacle for the acceptance of GDMS seems to be the problem of interferences. With respect to the experiences made with ICP-MS, this is difficult to understand. ICP-MS has nearly the same problems with interferences or more when the chemical pretreatment in preparing the analyte solution is considered. ICP-MS also uses a considerably higher Ar gas flow. Several years ago, people said that ICP-MS would not have a future because of the inherent problems with interferences. However, nowadays the same people have learned to live with these problems and to circumvent them in many cases. There is really no reason why the development should go a different way in case of GDMS.

The appearance of interferences is inevitably connected with the use of a working gas for the discharge. Interferences may be caused by

(1) isobaric atoms, e.g. $^{40}Ar^{+} \approx {}^{40}Ca^{+}$ and $^{92}Zr^{+} \approx {}^{92}Mo^{+}$,

(2) multiple charged atoms, e.g. $^{56}Fe^{2+} \approx {}^{28}Si^{+}$,

(3) molecular ions with atoms of the working gas, e.g. $^{40}Ar^{16}O^{+} \approx {}^{56}Fe^{+}$ and $^{56}Fe^{40}Ar^{+} \approx {}^{96}Mo^{+}$, and

(4) molecular ions resulting from residual contaminants, e.g. $^{14}N^{16}O^{1}H^{+} \approx {}^{31}P^{+}$.

Higher resolution reduces the problem of interferences but cannot totally avoid it. If the mass differences which must be resolved in case of interferences by hydride and argide ions are considered, it becomes obvious that in the case of the hydrides no resolution is possible above a certain threshold value and that in case of the argides interferences inside a certain interval around 100 Da cannot be resolved even with high resolution systems.

There are several effective ways of checking for the influence of interferences and how to avoid such influences in many cases. Of basic importance is the layout of the ion source and the ion extraction. This should be carried out from the hot inner zone of the plasma avoiding molecules from low temperature regions. Here the criterion is that the Debye-length of the plasma in the aperture region is small in comparison to the aperture diameter. Layout of ion source, ion extraction and ion transfer system may have a dramatic influence on the operation of the system, changing the ratio of interference to analyte ions by more than six orders of magnitude[2].

An initial check for interferences is possible by analyzing the isotopic pattern. In terms of data processing this can be realized by choosing the isotope which gives the lowest concentration value after correction for the abundance. More indicative is the check by a calibration procedure in which high y-axis intercepts of the extrapolated calibration line are a significant indication of an interference.

A very good, though time consuming, technique is the complementary use of alternative working gases[7]. A typical example is the mass area of Mo in an analysis of Fe which in case of Ar does not show the isotopic pattern of Mo, but corresponds well to the pattern of the matrix Fe[8]. This shows that the spectrum is dominated by the $FeAr^+$ molecules. But with Ne instead of Ar, the mass region in question will not be disturbed because the neides appear about 20 Da lower. This means that comparison of the spectra obtained with different working gases gives a lot of information concerning interferences and serves to enable analysis in many cases. Another important example is the determination of Fe in Al by its main isotope which is not possible with Ar due to the interference by $^{40}Ar^{16}O^+$, whereas a reliable determination of 1 ppm Fe is no problem with Ne as the working gas of the discharge.

Klingler et al.[9] have recently published a suppression of interferences by application of a pulse technique. The discharge is triggered by the power supply for a certain period. At the beginning of the afterglow after the discharge is extinguished, the intensity profile of sputtered species shows a sharp rise while the intensity of gas and hydride species almost vanishs, so that this afterglow phase shows best conditions for an analytical determination free of interferences. The determination of Si in Fe is normally impeded with low resolution by the interference of molecules like COH^+ or N_2H^+; however the spectrum which is obtained by gating for the afterglow phase enables a direct evaluation for the Si isotopes reproducing their abundance correct to within 1%.

A very sophisticated technique to obtain detailed information on interferences has been applied by King et al[10]. For a study of collision induced dissociation (CID), they used a triple quadrupole system in which a parent ion can be selected in a first quadrupole. The second quadrupole - rf only - serves as collision chamber and the third quadrupole selects the daughter ion, correspondig to MS-MS techniques in organic applications. In analysis of Cu for instance, the daughter spectrum of mass 81 revealed that this was not only caused by Ar_2H^+ but also by CuH_2O^+. Another example is the parent spectrum to mass 56 in a steel analysis which indicates that Fe^+ ions arise from $FeOH^+$, $FeAr^+$ and Fe_2^+. This principle has been applied by Duckworth et al.[11] to a low resolution system by introducing an rf-only quadrupole as collision cell in front of the analyzing quadrupole. A selective removal of problematic isobaric interferences seems possible.

6 OPTIMIZATION

The main operational parameters of a gas discharge are burning voltage, discharge current or power and gas pressure of which two can be chosen independently. The choice of operational parameters may strongly influence the analytical result so that parameter optimization can be very helpful to improve analytical performance and should be performed carefully for each type of sample. For the use of Ne and Ar in complementary analyses, it has been shown[8] that nearly identical performance can be obtained by choosing the pressure for Ne adequately higher than for Ar.

Another type of optimization refers only to the low resolution technique. Here application of a variable bias potential for all components of the ion optical system enables a representation of the ion kinetic energy distribution by simple variation of this bias potential. The resulting ion energy characteristic (IEC) offers the possibility of optimizing the ion transfer with respect to the mass[12]. Again, different settings are required in case of different working gases.

GDMS offers many possibilities for improving its performance in analytical procedures for any dedicated task. Utilization is only possible if the equipment offers the necessary variation range for the operational parameters.

7 ACCURACY

The primary question concerning precision and accuracy in case of a destructive analysis technique must be if we are justified in speaking of the same sample in procedure repetition. The question is really complex because there are many factors affecting accuracy. The first is the question of homogeneity or rather microinhomogeneities. These determine what amount of a sample must be analyzed to get a result which is representative for the solid as a whole. For instance, the determination of Si in Al is difficult (as already mentioned), because we may have precipitates at the grain boundaries. Furthermore, preparing the analyte sample may give rise to systematic errors. A very serious question is that of standards. An advantage of GDMS is that we can realize detection limits in the ppb region; but where are the corresponding standards, and are the present standards well enough characterized? For instance the analysis of a set of Cu standard samples[13] resulted in different sensitivities for a high alloy group and a low alloy group. However, this cannot be attributed to GDMS errors; it must be attributed to different standardization procedures. We should always be careful to discuss the question of accuracy only in terms of the spectrochemical determination method; in most cases, the other steps forming a procedure on the basis of the method are much more influential. For GDMS, one should always be aware that we can never be absolutely sure that really no interference is contributing to the analytical signal. In a very strict sense, we cannot interpret the determined value as a content but only as an upper limit of a content.

Keeping in mind that in most cases the question of accuracy is determined by the fact that there are no suitable standards or that the standards at hand are not well enough certified, the possibility of a semiquantitative analysis gains much interest. The question is: What can we do if we have no standards? A simple answer is to use the determined ion ratios directly without applying a sensitivity factor. In this case, the accuracy depends of course on the variation range of the sensitivity factors for the given matrix. A further possible answer is to use averages of sensitivities which have been determined by standards with a different matrix. In this case of a semiquantitative analysis, the accuracy depends on the influence by matrix effects. For GDMS, it is expected that there are no severe influences from matrix effects, and in general this expectation has been confirmed. If there are sensitivity variations, we must be aware that they may be caused also by operational differences of the equipment.

The lack of standards in the sub-microtrace region raises the question of what accuracy may be achieved in semiquantitative procedures. In general the variation range of elemental sensitivities is within one order of magnitude around unity. This means that a semiquantitative evaluation by ion beam ratios without sensitivity correction will result in an accuracy corresponding to a factor of about 3 over all. At first glance, this does not look convincing; but it is convincing if no standards are at hand, or standards at hand are not significantly better certified as is usually the case in the sub-microtrace region.

A second order approach for a semiquantitative analysis is to apply average RSF values which have been obtained with identical equipment for different matrices. Sanderson et al.[14] have published results obtained for the analysis of Mo with average RSF values in comparison to certified contents. These demonstrate in particular that the deviation between these is much lower than should be expected according to the variation range of sensitivities.

A further interesting aspect can be seen from investigations concerning the dependence of sensitivities on the gas pressure in low resolution GDMS which have been performed for steel standards[8]. There are two distinct regions. In a low pressure region, the variation range of sensitivities goes from 0.3 to 3.0 corresponding to the variation range of elemental sensitivities in GDMS. In a high pressure region, sensitivities are in general lower but the variation range is considerably smaller. The low pressure region with the higher sensitivities will result in lower detection limits. However, the high pressure region with the lower variation range of the sensitivities offers the chance to realize higher accuracies in case of analysis without standards.

A general warning should be included here concerning the transferability of sensitivity factors which have been determined under not absolutely identical conditions. Sensitivity factors determined with a multiplier as detector show a strong correlation to the atomic radius of the element[8]. However, it would be dangerous to apply these factors to an analysis with analog registration by a Faraday cup which shows totally different behaviour. The correlation to the atomic radius is only due to the varying sensitivity of the multiplier.

8 SPECIAL APPLICATIONS

As far as bulk analysis is concerned, GDMS is an established method. Furthermore, it is becoming more and more promising for new types of application which will be useful to draw a greater attention to this analytical tool.

The application of GDMS has been restricted so far to conducting or semi-conducting solids. In order to surmount this obstacle, Duckworth et al.[15] have recently tried to use an rf source instead of the dc source in a low resolution system. Direct analysis of non-conducting materials is also enabled by a special source offered for the VG 9000[16].

A performance study for the microchemical application of GDMS has been made with the determination of Pt and Ir[17]. Droplets of an analyte solution are pipetted to the surface of a Cu substrate. True cementation serves to prepare the sample for GDMS analysis of the residue on the substrate. The signal of an isotope representing the element to be determined is monitored as well as an internal standard. From the resulting single ion monitoring profiles the minimum detectable quantity was calculated to be clearly below 1 pg for both elements. In this type of experiment, a useful ion yield in the order of 10^{-5} has been achieved which is considerably above the corresponding value for SNMS. It should be mentioned that the experiment was performed with a pin-shaped cathode, while the corresponding experiment with a disk-shaped cathode resulted in a useful yield more than one order of magnitude lower.

The disk-shaped cathode with a Grimm-type discharge is the method of choice for surface and in-depth analysis for which sputtering is a good method of atomization, because only atoms of a small surface region with a certain information depth are emitted. For depth-profiling with satisfactory depth resolution, it is of course necessary to realize plane crater profiles so that the intensity-to-time profile can be converted to a concentration-to-depth profile. Here we have once more an optimization problem, as there is a certain burning voltage for each type of sample at which this can be realized. For lower or higher voltages, the crater profile will show curvature - convex or concave - due to distortions of the electrical field.

In a systematic study of the performance of GDMS for the analytical characterization of surface layers[18], a depth resolution down to about 10 nm has been realized. In this case, Ne was used as the discharge working gas in order to keep the sputter rates lower than is usual with Ar. On the other hand, depth profiling of a 10 μm double layer on a substrate was possible in 5 minutes. This demonstrates that GDMS is particularly useful for analyses in which the high depth resolution of SNMS is not necessary. This is the case for the majority of technical surface layers, such as used for passivation for instance, of which the thickness goes up into the μm region.

GDMS has also been applied to organic analysis. With a special version of a GDMS ion source for direct atmospheric sampling[19], a minimum detectable quantity of 400 fg corresponding to 1 ppt has been achieved for the determination of trinitrotoluene in preliminary tests. With another special ion source, it has been shown[20] that glow discharge ionization acts similarly to rapid desorption chemical ionization. This means that the technique is in particular useful for "difficult" molecules in the sense of organic MS. The mechanism of ion generation is a rapid sputter heating of the probe tip with subsequent evaporation.

Finally another interesting source should be mentioned which has been used by Kim et al[21]. Basically, it is a special version of a Grimm type glow discharge lamp or source in which a gas jet is directed towards the sample surface in order to realize higher sputter efficiency. A similar approach has been reported by Horlick[22] who equipped a commercial ICP-MS system with two GD sources with pin geometry and disc geometry, so that his instrument can now be used for GDMS as well as for ICP-MS. This follows the concept of Jakubowski et al. who used identical MS equipment for a low resolution GDMS system[2] and a laboratory ICP-MS system[12]. Obviously, there is a certain consensus about a good prognosis for a combination instrument enabling ICP-MS and GDMS with identical MS equipment in a low resolution version and perhaps in a high resolution version, too.

REFERENCES

1. K.R. Hess, W.W. Harrison, Anal. Chem., 1988, 60, 691.
2. N. Jakubowski, D. Stuewer, G. Toelg, Int. J. Mass Spectrom. Ion Proc., 1986, 71, 183.
3. A.P. Mykytiuk, P. Semeniuk, S. Berman, Spectrochim. Acta Rev., 1990, 13, 1.

4. L.F. Vassamillet, J. Anal. At. Spectrosc.,1989, 4, 451.
5. G. Kudermann, Fresenius Z. Anal. Chem., 1988, 331, 697.
6. N. Jakubowski, D. Stuewer, W. Vieth, Anal. Chem., 1987, 59, 1825.
7. N. Jakubowski, D. Stuewer, W. Vieth, Fresenius Z. Anal. Chem., 1988, 331, 145.
8. N. Jakubowski, D. Stuewer, Fresenius Z. Anal. Chem., 1989, 335, 680.
9. J.A. Klingler, P.J. Savickas, W.W. Harrison, J. Am. Soc. Mass Spectrom., 1990, 1, 138.
10. F.L.King, W.W. Harrison, Int. J. Mass Spectrom. Ion Proc., 1989, 89, 171.
11. D.C. Duckworth, R.K. Marcus, 1990 Winter Conf. Plasma Spectrochem., St. Petersburg FA, paper W2.
12. N. Jakubowski, B.J. Raeymaekers, J.A.C. Broekaert, D. Stuewer, Spectrochim. Acta, 1989, 44B, 219.
13. N. Jakubowski, D. Stuewer, unpublished.
14. N.E. Sanderson, P. Charalambous, D.J. Hall, R. Brown, J. Res. NBS, 1988, 93, 426.
15. D.C. Duckworth, R.K. Marcus, Anal. Chem., 1989, 61, 1879.
16. U. Greb, J. Clark, G. Ronan, 2. Int. Conf. Plasma Source Mass Spectrom., Durham UK, 1990.
17. N. Jakubowski, D. Stuewer, Spectrochim. Acta B, in press.
18. N. Jakubowski, paper in preparation.
19. S.A. McLuckey, G.L. Glish, K.G. Asano, B.C. Grant, Anal. Chem., 1989, 60, 2220.
20. R. Mason, D. Milton, Int. J. Mass Spectrom. Ion Proc., 1989, 91, 209.
21. H.J. Kim, E.H.Piepmeier, G.L. Beck, G.G. Brumbaugh, O.T. Farmer, Anal. Chem., 1990, 62, 639.
22. G. Horlick, 1990 Winter Conf. Plasma Spectrochem., St. Petersburg FA, paper PL5.

Direct Analysis of Semiconductor Grade Reagents by ICP-MS

Amanda Walsh, Donald Potter, Edward McCurdy, and Robert C. Hutton

VG ELEMENTAL LTD, ION PATH, ROAD THREE, WINSFORD, CHESHIRE CW7 3BX, UK

1. INTRODUCTION

The excellent low detection power afforded by ICP-MS has been used to great advantage in many analytical applications. None more so than in the analysis and control of trace element impurities in the reagents used in semiconductor processing. The trace element content of these reagents, used as wash solutions, and etchants etc is of critical importance in determining the final quality and performance of microelectronic devices. Analysis of these reagents is particularly demanding, requiring rigorous control over the laboratory environment and also over the analytical protocols employed in the measurements; thus specialised techniques must be used if accurate quantitation is to be maintained at pgg^{-1} levels. As the industry moves towards the production of more powerful mega chip devices however, the analytical requirements become more demanding and the limitations of current quadrupole based ICP-MS become obvious. For the future therefore, instrumentation capable of circumventing such limitations is required. To this end, high resolution ICP-MS offers to take the technology one step further on both detection power and accuracy of quantitation. This paper discusses the performance of ICP-MS in this particular application and compares and contrasts both quadrupole and high resolution instrumentation illustrating how both can be used effectively.

2. EXPERIMENTAL

Quadrupole ICP-MS measurements were performed on a VG PlasmaQuad PQ2 Plus Turbo series instrument. The sample introduction system was replaced by one which was specially designed to be impervious to attack from those reagents and also to minimise contamination. A summary of this system is given in Table 1.

High Resolution ICP-MS measurements were performed on a VG PlasmaTrace instrument. This is a double focusing mass spectrometer capable of operating at resolutions (M/ M) of up to 10,000. A schematic of the instrument is shown on Figure 1. A detailed description of the operational aspects of this instrument can be found in Ref [1]. The sample introduction system was the same as that used for quadrupole ICP-MS.

Table 1

Sample Introduction System for High Purity Reagents

ICP	Quartz Torch
SPRAY CHAMBER	High Purity fused quartz double pass design or PTFE fabricated chamber, Double pass design. Both systems were water cooled and maintained at 5°C.
NEBULIZER	V-Groove type of de Galan design. Fabricated from KEL-F
SAMPLE CONE	Platinum, 1.00mm orifice. Nicone design.[7]
SKIMMER CONE	Nickel or Platinum. Nicone design.[7]

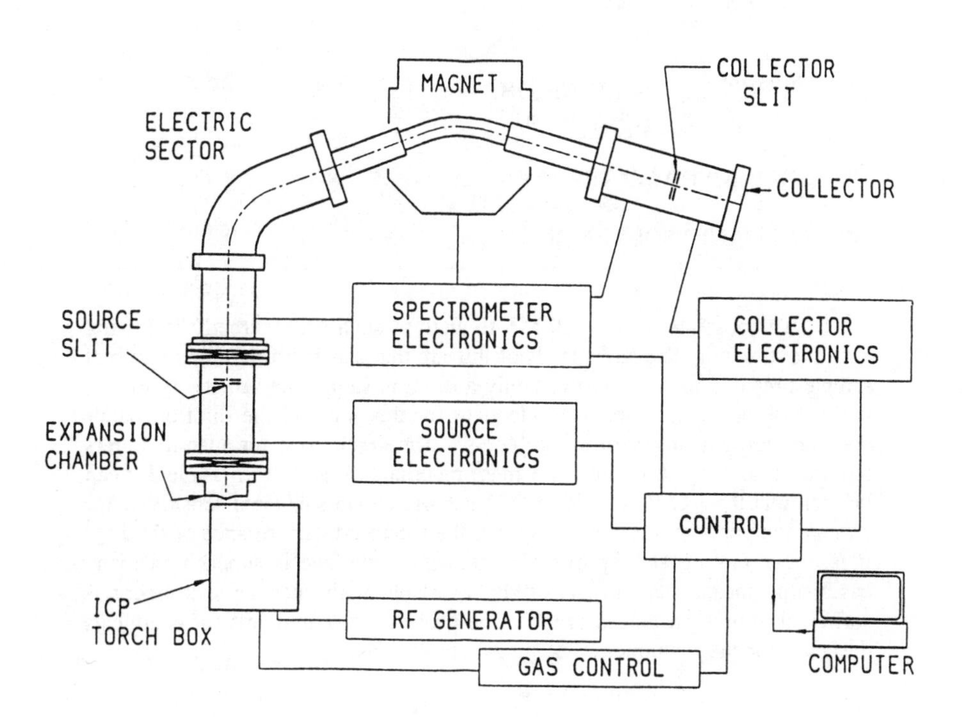

Figure 1 High Resolution ICP-MS Schematic Diagram

3. RESULTS/DISCUSSIONS

The reagents which require analysis in the microelectronics industry are many and varied. Table 2 summarises many of these regents. The methods used currently typically employ an evaporative preconcentration step followed by an analytical finish with graphite furnace AA. The reasons for this are, that it can give a preconcentration factor of ~50x thus enhancing detection power. [2] Also, evaporative preconcentration removes the majority of the reagent matrix making standardisation and analyses more straightforward. However, this method is time consuming and requires great care to minimise sample contamination. Also, preferential loss of volatile elements such as As, Se, Zn, and particularly B from, for example, HF can degrade accuracy.

Table 2

	Reagents Requiring Analysis	
1.	ACIDS	HNO_3, HF, H_2SO_4 HCl, H_3PO_4, H_2O
2.	BASES	NH_4F, NH_4OH, $(CH_3)_4$ N OH
3.	DEIONISED WATER	
4.	ORGANIC SOLVENTS	

It is possible, using ICP-MS to analyse such acids directly following a simple dilution. The advantages of this are that it is rapid, requires minimal sample preparation, and involves only a dilution step. The sensitivity of ICP-MS is however more than adequate for this job and the dilution factor does not present any serious problems. The semiconductor manufacturers require certain detection limits for these materials [3] and from Table 3 it can be seen that the VG PlasmaQuad PQ2 has more than sufficient sensitivity for current analytical requirements. Whilst the enhanced performance of the PQ2 Plus, incorporating an improved ion transport interface, is superior in many cases, this means that the sensitivity available with such an instrument is sufficient in most instances even for the projected requirements for chemicals designed for the manufacture of megabyte devices.

Table 3

Comparison of PQ2 and PQ2 Plus Detection Limits of S.E.M.I specification elements (all values in ppb)

Element	PQ2	PQ2+	S.E.M.I.-Spec [2]
Aluminium	0.06	0.01	500
Antimony	0.005	0.001	5
Arsenic	0.05	0.002	5
Boron	0.05	0.05	10
Cadmium	0.01	0.002	500
Calcium	5	0.5	1000
Chromium	0.04	0.005	500
Cobalt	0.01	0.0005	500
Copper	0.03	0.004	100
Gallium	0.03	0.0007	50
Germanium	0.08	0.001	500
Gold	0.08	0.008	500
Iron	0.1	0.1	200
Lead	0.05	0.0004	100
Lithium	0.05	0.001	1000
Magnesium	0.05	0.002	1000
Manganese	0.01	0.01	500
Nickel	0.05	0.01	100
Potassium	5	0.5	1000
Silver	0.01	0.001	500
Strontium	0.01	0.0005	500
Tin	0.005	0.0005	500
Zinc	0.01	0.006	500

Analysis of Acid Reagents

The reagents vary [4] from simple to analyse hydrogen peroxide to the most difficult of the reagents the acids, and in particular, phosphoric and sulphuric acid which being viscous are difficult to analyse without dilution. Each can attack and seriously degrade the conventional nickel samplers used in ICP-MS. However, platinum samplers are impervious to such attack and concentrations up to 20% v/v H_2SO_4 can be run directly with no problems. Another reagent which causes difficulty is ammonium hydroxide. Whilst it is not particularly
chemically aggressive and does not attack metal, the presence of large amounts of NH_4OH in an argon ICP may, under certain circumstances cause the ICP to extinguish. The crystal controlled matching network used with the PlasmaQuad is capable of responding rapidly to this situation and even 1:1 diluted ammonia can be run with no ICP problems. Table 4 illustrates detection limits for H_2SO_4, H_3PO_4 and NH_4OH following 1:10 dilution, and for NH_4OH, following 1:5 dilution.

The figures vary little from aqueous detection limits, illustrating the efficacy of this approach.

Table 4

	Detection Limits		
Element	**Sulphuric Acid ng g^{-1}**	**Ammonium Hydroxide ng ml^{-1}**	**Phosphoric Acid ng g^{-1}**
Be	0.05	0.019	0.4
Mg	0.02	0.001	0.7
Cr	0.01	0.003	---
Mn	0.01	0.002	---
Co	0.015	0.010	0.16
Ni	0.06	-----	0.55
Cu	0.04	0.010	---
Zn	0.04	0.04	0.65
Ga	0.02	0.006	0.16
Sr	0.01	0.006	0.032
Y	0.003	0.001	0.015
Mo	0.29		0.083
Zr	0.07	-----	0.065
Ay	0.025	0.010	----
Cd	0.05	0.070	0.33
In	0.01	-----	0.05
Sn	0.05	-----	0.04
Sb	0.02	0.015	-----
Ba	0.04	-----	-----
Ce	0.01	-----	0.02
Au	0.013	0.010	-----
Tl	0.009	-----	0.035
Pb	0.02	0.015	0.42
Bi	0.009	0.007	0.05
Th	0.008	0.005	-----

Quantitation

Whilst the instrument used may have the detection capabilities for sub ng g^{-1} measurement, quantitation at such levels presents other problems. Typically, external calibration is employed in ICP-MS, which is rapid and convenient however, with analysis of high purity reagents it is difficult tọ use internal standards for fear of contaminating the sample. Also matrix responses may vary between acid types causing anomalous responses. It has been found [5], that for quantitation of these reagents, the method of standard addition provides more accurate analytical results. The advantages are that it is matrix specific, gives good accuracy when the isotopes used for measurement do not have interferences, and does not contaminate the sample. It does not, however, compensate for spectral interferences. Analysis of sulphuric acid and hydrogen peroxide using the method of standard additions is given in Table 5 and 6. It can be seen that the values obtained are in almost all cases less than the manufacturer's specification. Values for Ca and Fe in H_2SO_4 were measured by external calibration since both elements suffer spectral interferences on all isotopes. Examples of linear calibration plots in 1:10 diluted H_2SO_4 (Figure 2) and H_3PO_4 (Figure 3) are shown. Even with the difficult matrix, excellent linearity was obtained, being generally yielding correlation coefficients better than 0.999.

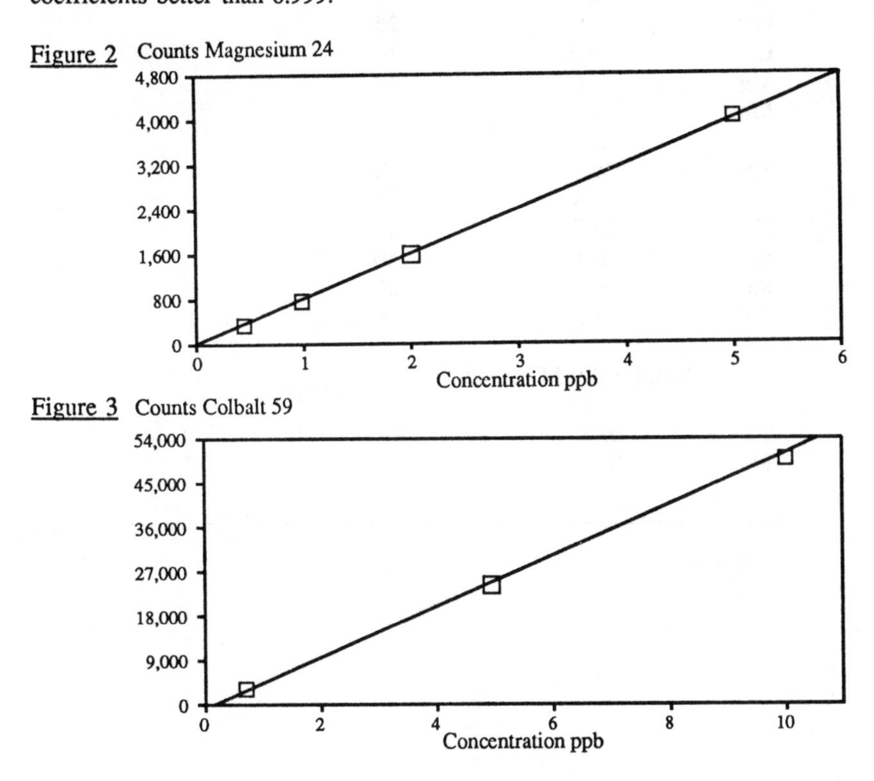

Figure 2

Figure 3

Table 5

Analysis of Sulphuric Acid

Element	Concentration ng g^{-1}	Manufacturers Max Specifications
As	1.9	2
Sb	0.5	2
Al	6.5	10
Ba	1.1	10
B	8.5	10
Cd	0.7	10
Ca	<20	25
Cr	11.7	10
Co	1.1	10
Cu	1.2	10
Ga	<0.02	10
Ge	2.2	10
Au	0.4	10
Fe*	11.9	25
Li	<0.02	10
Mg	1.7	10
Mn	0.7	10
Ni	2.7	10
K*	19	10
Ag	16.4	20
Na	11.2	20
Sr	0.4	10
Sn	9.0	20
Zn*	4.7	10
	Additionally Measured	
Tl	0.08	
Pb	0.25	

Table 6

Analysis of Hydrogen Peroxide

Element	Concentration ug l^{-1}	Element	Concentration ug l^{-1}
Be	0.01	Ag	0.25
Mg	3	Cd	0.15
Cr	1.5	Sn	2
Mn	0.7	Sb	0.04
Co	0.06	Ba	4.5
Cu	0.7	Ce	2
Zn	2.9	Ta	10
Ga	0.3	Au	0.085
Sr	0.12	Pb	0.65
Y	0.0015	Bi	0.05
Nb	0.7	Th	0.5

Spectral Interferences

Of the elements which require to be analysed however, some suffer spectral interferences from polyatomic ion species. For example, ^{56}Fe is overlapped by ArO^+ causing detection limits to be ~ 10 ng g^{-1} in acid. This is inadequate for current and for future performances and alternative techniques must be sought for this critical analysis. Similarly once quantitation is required at low pg g^{-1} levels, the presence of small spectral interferences, primarily on the transition metals may be a problem. At higher measurement levels however, some of these interferences are barely discernible above background. And there are a variety of techniques which potentially could be employed to overcome these,eg. ETV-ICP-MS could in theory be used. It has the capability to remove the matrix prior to analysis whilst there is also the advantage of enhanced sensitivity. ArO could therefore be separated from Fe in the thermal cycle of the ETV, and the Fe left free for analysis [6].
Figure 4 illustrates an ETV calibration for Fe showing a detection limit of < 20pg ml^{-1}. This is certainly useful for Fe analysis in such samples. This approach is not applicable to all problematic elements however, and the possible solution is the use of a high resolution ICP-MS instrument.

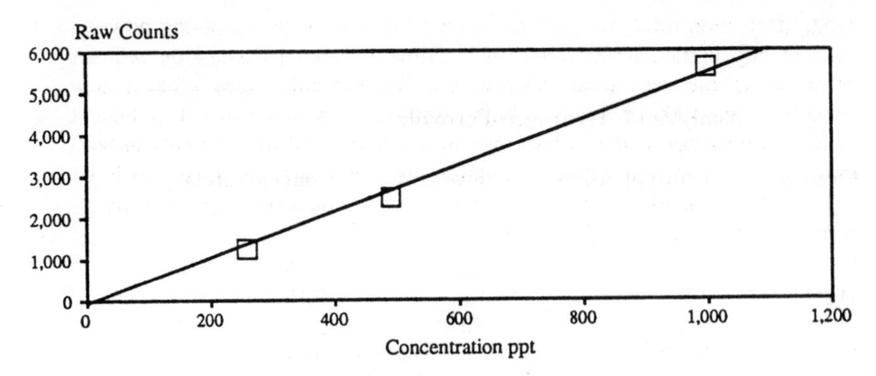

Figure 4 Iron Calibration in 2% Nitric Acid

High Resolution ICP-MS

The use of a high performance, high resolution mass spectrometer in this application yields several advantages for the analysis. The spectrometer is capable of operating up to resolving powers of about 10,000. This is sufficient to separate many of the common acid based spectral interferences as shown in Table 7 and hence several isotopes which were previously inaccessible can now be used for analysis. Another advantage of the configuration of instrument is, that the level of photon derived background noise is significantly lower than with the quadrupole instrument being typically <~0.1 cps. This is a direct result of both the radial geometry and the presence of source and collector slits in the mass spectrometer Hence the expected detection limits should be significantly better than with the quadrupole based on background considerations alone.

Table 7

Resolving Power of VG PlasmaTrace

Element	Isotope	Interferant	Required Resolution
Si	28	N_2	957
S	32	O_2	1800
Fe	56	ArO	2501
Tl	48	SO	2518
Ti	47	PO	2776
V	51	ClO	2572
Cr	52	ArC	2374
Zn	64	SO_3/S_2	1951/4262
Cu	63	PO_2	1851
As	75	ArCl	7771
Se	77	ArCl	9181

Detection limits will of course be a function of the resolving power used in any particular measurement. Some reduced transmission will be observed as the resolution is increased for example from ~800 (low resolution) to ~4000 (medium to high resolution). A factor of ~10 is typical for this comparison if detection limits in sulphuric acid for elements in both low and (800) and high (4000) resolution are given in Table 8. It should be noted that the values quoted refer to the original acid and are given in pg g^{-1} after incorporating a 10x dilution factor.

Table 8

High Resolution ICP-MS Detection Limits For Sulphuric Acid (96%)

Element	Detection Limit	
	Low Resolution	High Resolution
Mn		18
Cu		3.6
Zn		1.2
Au	0.025	
Th	0.085	
U	0.11	

Values in pg g^{-1} (ppt)

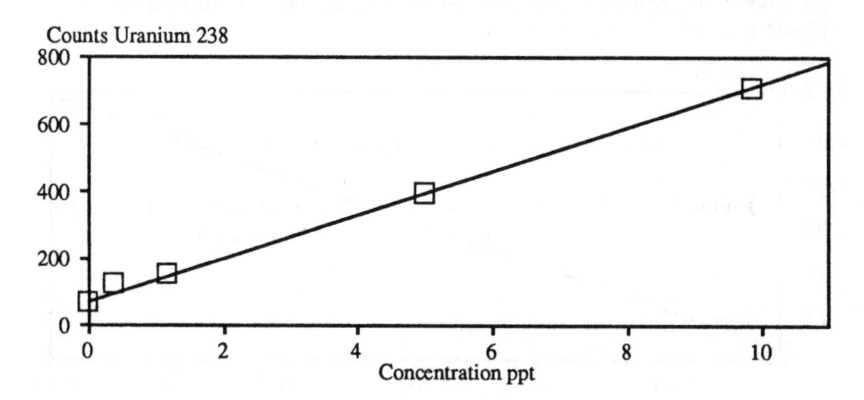

Concentration of U in 10% Sulphuric Acid = 1ppt
(Determined by standard addition)

Figure 5 Six Point Calibration Graph, U in 10% Sulphuric Acid

A six point calibration curve for uranium in 10% v/v H_2SO_4 is illustrated in Figure 5. The fit of the line is excellent especially considering the ultra trace levels used is these experiments. To illustrate the sensitivity of the instrument further, Figure 6 shows a spectral display from 500 ppq (0.5 ppt) of U in 10% v/v H_2SO_4. It can be clearly seen that the detection limits obtained are in fact not theoretical.

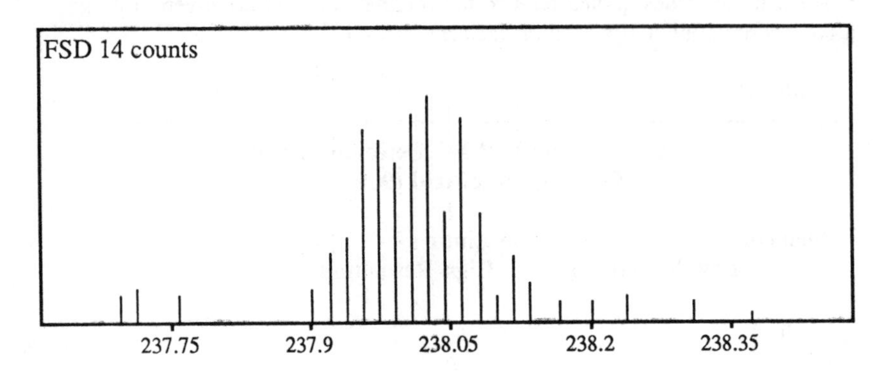

Figure 6 500ppq U addition in 10% Sulphuric Acid

The performance of the instrument in high (~3500) resolution is shown for ^{63}Cu in 10% v/v H_2SO_4 in Figure 7. Again excellent linearity is obtained on a six point calibration to 1 ppb and the separation of the instrument illustrated for isotope of ^{55}Mn separated from ArNH also at nominal mass 55 on Figure 8.

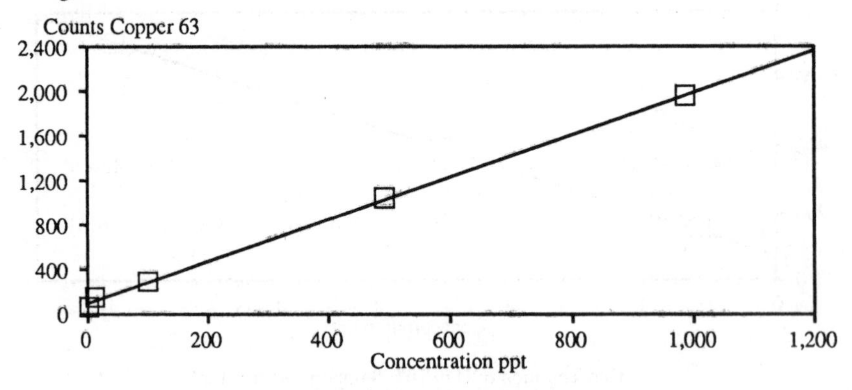

Figure 7 Six Point Calibration Graph, Cu in 10% Sulphuric Acid

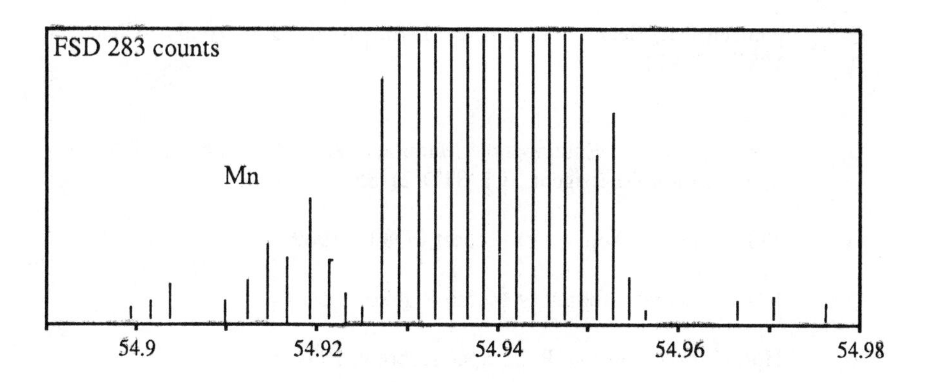

Figure 8 10ppt Mn Addition in 10% Sulphuric Acid

4. CONCLUSIONS

The excellent detection power of ICP-MS makes it the technique of choice for the demanding analysis of the reagents used in semiconductor production. Reagents such as ammonia, hydrogen peroxide, sulphuric and phosphoric acids can be analysed using quadrupole ICP-MS to the levels currently required for the industry. Although the technique is not applicable to all important elements there are alternative strategies which can be employed.

As megachip technology develops however, the requirements for lower detection limits becomes stronger. The parallel development of high resolution ICP-MS, incorporating double focusing technology new provides the capability not only for measurement at ppq (fg ml^{-1}) levels but also the ability to identify absolutely elemental impurities even when they may be present at the same nominal mass as a low level spectral interference. It is envisaged that as both technologies develop in parallel, that high resolution ICP-MS will become the definitive technique for trace element analysis of advanced materials, whilst quadrupole based technology will develop as the technique for quality control of production reagents.

5. REFERENCES

1. Bradshaw, N., Hall, E.F.H., Sanderson, N.E., J.Analyt. Atom. Spectrom., 1989, 4, 801.

2. Paulsen P.J., Beary E.S., Bushee, D.S., Moody, J.R., Analyt. Chem., 1988, 60, 971.

3. Semiconductor Equipment Manufacturers Institute (S.E.M.I.) Specification for reagents, CI, STD 21-85.

4. VG Elemental Application Report PQ801, 1989.

5. VG Elemental Laboratory Memo PQ590, 1990.

6. Hulmeston.P, Hutton. R.C., Spectroscopy, 1991, 6, 35.

7. US Patent No. 4 760 253, 26.7.88.

Analysis of Thermal Waters by ICP-MS

E. Veldeman[1], L. Van't dack[1], R. Gijbels[1], M. Campbell[2], F. Vanhaecke[2], H. Vanhoe[2], and C. Vandecasteele[2]

[1] UNIVERSITY OF ANTWERP (U.I.A.), UNIVERSITEITSPLEIN 1, B-2610 ANTWERP, BELGIUM
[2] UNIVERSITY OF GHENT (R.U.G.), PROEFTUINSTRAAT 86, B-9000 GHENT, BELGIUM

1 INTRODUCTION

In the course of a broad geochemical study on the behaviour of trace elements during water-rock interaction, a series of hydrothermal waters were sampled in the Rila-Rodope region of southwest Bulgaria. In this district, N_2-charged waters of low salinity (0.2-0.8 g/l) flow out at temperatures up to 96°C. These waters are associated with highly fissured granites and metamorphic silicate rocks. A whole range of trace elements (Li, B, Ga, Ge, Rb, Cs, W,..) is leached from the granites into the hydrothermal solution. These elements can be used as a criterion for the characterization of waters originating from a granitic environment[1]. Moreover, the trace element content of a thermal water can yield valuable information on the reservoir conditions, in addition to the conclusions drawn from the major element chemistry of the investigated water. Both the broad trace element pattern and the low mineralization make these waters very well suited for analysis by the ICP-MS multi-element technique. The water samples could be analysed directly without any dilution or matrix separation prior to the ICP-MS analysis.

Further, the thermal water samples were also analysed by Instrumental Neutron Activation Analysis (INAA) after freeze-drying. This technique has successfully been applied to the analysis of thermal waters from the Vosges area (France) by Blommaert (1983). ICP-MS, unlike INAA, is a fast technique which yields relatively simple spectra and requires minimal sample preparation. Results from both analytical methods are compared in this paper.

ICP-MS was used for the analysis of natural waters by, amongst others,: Beauchemin and co-workers[2-4], Garbarino and Taylor[5], Hall et al.[6], and Henshaw et al.[7]. In particular, Sansoni et al.[8] compared the performance of ICP-MS with that of ICP-OES, ICP-AFS and FAAS for the multi-element analysis of natural waters from the Fichtelgebirge granite region (Germany).

2 EXPERIMENTAL

ICP-MS

Instrumentation. The ICP mass spectrometer was a VG PlasmaQuad (VG Elemental, Winsford, UK), used in its standard configuration, with a Meinhard nebulizer and a spray chamber made of borosilicate glass. The operating conditions are summarized in Table 1.

Table 1. Operating conditions

Plasma		*Mass spectrometer*	
Power, kW	1.35	Expansion stage, mbar	2.7
Nebulizer gas flow, l min^{-1}	0.728	Intermediate stage, mbar	$1.1x10^{-4}$
Auxiliary flow, l min^{-1}	1.1	Analyser stage, mbar	$2.0x10^{-6}$
Plasma gas flow, l min^{-1}	14		
Sample uptake rate(pumped), ml min^{-1}	0.9	*Scan conditions*	
Sample depth above load coil, mm	10	Mass range, u	6-240
Spray chamber cooled to, °C	10	Number of scan sweeps	150
		Dwell time, µs	320
		Number of channels	2048

Solutions. Multi-element standards were prepared in 0.14 M nitric acid, by dilution of 1 g/l stock solutions of the individual elements. The thermal water samples were filtered (0.4 µm, Nuclepore) and acidified (0.2% v/v HNO_3) immediately after sampling. After transport to the laboratory, the water samples were analysed as such, without preliminary dilution or matrix separation. Both standard and sample solutions were spiked with 100 µg/l In as an internal standard.

Measurements. Since the elements of interest cover the whole mass range, the ICP-MS was used in the full spectrum scanning mode. The scan conditions are summarized in Table 1. Each measurement lasted for ca. 98 s. Each solution was measured three times. In order to avoid memory effects, the sample introduction system was rinsed with 0.14 M nitric acid for several minutes between the analysis of the different sample solutions.

INAA

After filtration (0.4 µm, Nuclepore), acidification to pH 1.5 and transport to the lab, 100 ml portions of the water were pipetted into polyethylene bags (Freshpack, UCB) and frozen in liquid nitrogen. Afterwards, the samples were freeze-dried in a Lyo 2 apparatus (Leybold-Heraeus). After complete evaporation of the water, the plastic bags with the sample residue were pelletized (Φ=12 mm, height 3-5 mm). Then, the samples were irradiated in the nuclear reactor Thetis (Ghent) at a flux of 10^{12} n.cm^{-2}.s^{-1}, together with a copper flux monitor. The analysis procedure comprises two different irradiations, followed by several measurements of the resulting γ-rays by a Ge(Li) detector. A detailed description of the applied procedure is reported by Blommaert[9].

3 RESULTS AND DISCUSSION

Matrix Effects and Interferences

Table 2 summarizes the results from ICP-MS analysis of a 4-fold, 2-fold and undiluted water sample.

The excellent agreement between the results from diluted and undiluted samples indicates that possible matrix effects, e.g. due to the presence of easily ionized elements, are adequately corrected for by using In as internal standard[10].

Table 3 presents the chemical characteristics of the water samples investigated. The hydrothermal waters are of the Na-SO_4-HCO_3 type. This suggests that the most important matrix interferences are expected to be due to Na-, S- and C-containing species, in addition to the background spectral features from the nitric acid aqueous solution[11]. However, in general, matrix interferences are only very minor. The significantly higher copper concentrations calculated by use of isotope 63 relative to isotope 65 (see Table 2) reveal the presence of a $^{40}Ar^{23}Na$ interference at m/z 63. In a similar way, the presence of a $^{32}S^{16}O_2$ and/or $^{32}S^{32}S$ interference at m/z 64 can be inferred from comparison of the Zn concentrations calculated using ^{64}Zn and ^{68}Zn isotopes. Due to the rather low chloride content of the investigated waters (2-30 mg/l), ClO and ArCl interferences at m/z 51/53 and 75/77 are very minor. Furthermore, no interference at m/z 52, due to $^{40}Ar^{12}C$, was detected. In spite of a rather high SiO_2 content of the high-temperature thermal waters, Si didn't hamper the determination of nickel ($^{28}Si^{16}O_2$ at m/z 60) nor zinc ($^{40}Ar^{28}Si$ at m/z 68) (see Table 2).

Table 2. Results (μg/l) obtained by ICP-MS for 26 elements in a selected thermal water sample (mean ± standard deviation of 3 measurements)

	4-fold diluted	2-fold diluted	not diluted
Li(7)	140 ± 7	143 ± 7	155 ± 5
B(11)	450 ± 16	446 ± 36	473 ± 28
Mg(24)	7.6 ± 0.8	6.7 ± 0.6	6.4 ± 0.5
Al(27)	91 ± 4	86 ± 4	85 ± 7
Ca(44)	5200 ± 300	5500 ± 500	5500 ± 300
Ti(47)	5.4 ± 0.3	5.8 ± 0.8	6.3 ± 0.7
V(51)	0.14 ± 0.08	0.12 ± 0.04	0.15 ± 0.03
Cr(52)	<0.20	<0.10	<0.05
Mn(55)	0.44 ± 0.08	0.48 ± 0.03	0.51 ± 0.06
Fe(57)	16 ± 3	12 ± 3	15 ± 5
Ni(60)	<0.9	<0.5	<0.23
Cu(63)	24 ± 3	19 ± 4	20 ± 4
Cu(65)	1.0 ± 0.7	1.1 ± 0.4	1.1 ± 0.2
Zn(64)	8 ± 1	7 ± 1	6.9 ± 0.6
Zn(68)	<0.5	0.38 ± 0.09	0.4 ± 0.2
Ga(69)	6.3 ± 0.4	6.1 ± 0.1	6.4 ± 0.2
Ge(72)	20 ± 1	20.3 ± 0.9	19.4 ± 0.5
As(75)	0.6 ± 0.3	0.8 ± 0.2	0.67 ± 0.09
Rb(85)	62 ± 2	63.3 ± 0.8	62.3 ± 0.9
Sr(88)	180 ± 8	182 ± 1	183 ± 3
Mo(98)	8.6 ± 0.8	9.1 ± 0.6	6 ± 2
Sn(120)	0.48 ± 0.08	0.4 ± 0.1	0.3 ± 0.1
Sb(121)	0.27 ± 0.08	0.32 ± 0.02	0.31 ± 0.03
Cs(133)	67 ± 3	70 ± 1	68 ± 2
Ba(138)	1.0 ± 0.1	0.85 ± 0.04	0.90 ± 0.02
W(182)	250 ± 12	258 ± 17	263 ± 11
Pb(208)	<0.24	<0.12	<0.06
U(238)	<0.008	<0.004	<0.002

Table 3. Chemical characteristics of Bulgarian thermal waters (mg/l)

pH : 8-10	SiO_2 : 20-120
ΣH_2CO_3 : 20-350	Na : 40-300
F : 1-15	K : 0.1-10
Cl : 2-30	Ca : 1-10
SO_4 : 10-300	Mg.: 0.001-0.1

Detection limits

Detection limits for ICP-MS and INAA are compared in Table 4, from which it is clear that ICP-MS detection limits are more uniform over the whole mass range relative to INAA detection limits. Furthermore, ICP-MS detection limits are often lower than those for INAA. Finally, only very few elements are *a priori* excluded for analysis by ICP-MS. However, for INAA some elements of interest do not form activation products with easily measurable gamma rays (e.g. Ge, Sn, Pb).

Comparison of ICP-MS results with those from INAA

The results from ICP-MS and INAA multi-element techniques were compared by use of linear regression. When no systematic errors are involved for both methods, the slope and the intercept of the regression line should not be significantly different from 1 and 0, respectively. However, this approach is open to serious theoretical objections. First, the line of regression is calculated on the assumption that the errors in the x-values are negligible. Further, it is assumed that the error in the y-values is constant. Both assumptions obviously are invalid in practice. Meanwhile, despite the theoretical objections, the applied approach yields useful information, provided that the following requirements are fulfilled[12]:

1. The more precise method is plotted on the x-axis.
2. At least 10 data points are involved.
3. The experimental points should cover the concentration range in a roughly uniform fashion.

Table 4. ICP-MS and INAA detection limits (μg/l) for 32 elements in a typical thermal water sample

A(Isotope)	ICP-MS	INAA
Li(7)	0.04	ND
B(11)	1.2	ND
Mg(25)	0.2	3000
Al(27)	0.1	20
Ca(44)	3	2000
Ti(47)	0.4	250
V(51)	0.03	0.8
Cr(52)	0.05	1
Mn(55)	0.04	7
Fe(57)	2	30
Co(59)	0.01	0.05
Ni(60)	0.2	60
Cu(65)	0.1	150
Zn(68)	0.13	3
Ga(69)	0.04	ND
Ge(72)	0.05	ND
As(75)	0.07	3
Se(77)	1	1
Rb(85)	0.03	2
Sr(88)	0.02	20
Mo(98)	0.02	2
Cd(111)	0.08	20
Sn(120)	0.05	ND
Sb(121)	0.04	0.1
Cs(133)	0.04	0.02
Ba(138)	0.01	90
La(139)	0.02	0.4
Ta(181)	0.002	0.08
W(182)	0.007	1
Pb(208)	0.06	ND
Bi(209)	0.003	0.6
U(238)	0.002	0.2

ND : Not Determined

In general, a satisfactory agreement between both analytical methods was found. The results obtained by ICP-MS and INAA for Rb and Cs in the thermal waters investigated are presented in Fig. 1.

Fig. 1. Results obtained by ICP-MS and INAA a) for Rb (μg/l) b) for Cs (μg/l) in the waters investigated.

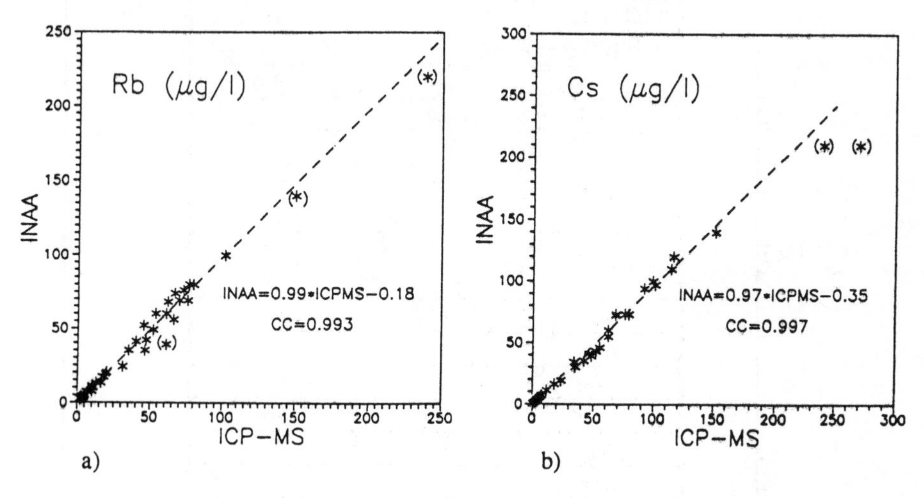

Fig. 2 compares the results for 13 elements (Li, Mg, Al, Mn, Ga, Ge, As, Rb, Sr, Mo, Sb, Cs, W) in a selected thermal water sample obtained by different analytical methods (INAA, Flame-AES, Flame-AAS, Graphite Furnace-AAS and DC-AES). In general, the agreement between the ICP-MS data and those from other techniques is satisfactory. The Al concentration measured by INAA, however, is significantly higher than the value from ICP-MS and Graphite Furnace-AAS. This may be due to the presence of particulate material (<0.4 μm), which possibly was not analysed by ICP-MS and Graphite Furnace-AAS. According to Hem[13], gibbsite crystals near 0.10 μm in diameter, for instance, have considerable physical and chemical stability.

4 CONCLUSIONS

This work has shown that ICP-MS is a powerful technique for the analysis of thermal waters originating from a granitic region. These waters display a broad trace element pattern and are weakly mineralised, such that few matrix effects and interferences are involved. Detection limits for most elements are on the 0.X to 0.00X μg/l level. Results from ICP-MS and INAA on these waters agree very well.

Fig. 2. Data obtained by different analytical methods for Li, Mg, Al, Ga, Ge, As, Rb, Sr, Mo, Sb, Cs and W in a typical thermal water sample.

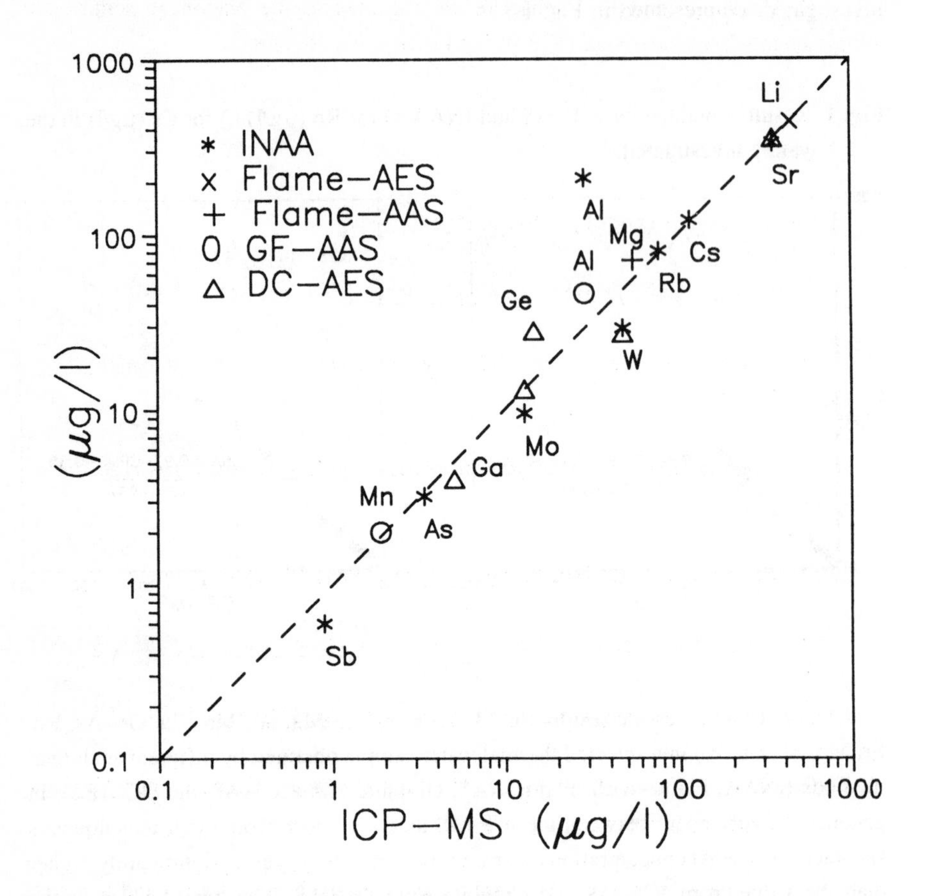

For the analysis of the investigated water samples, ICP-MS offers several advantages over INAA. These include:

1. Wider multi-element analysis capability.
2. Fast mode of operation.
3. Simple spectra.
4. Small sample size.
5. Lower detection limits for many elements.
6. Limited sample preparation.

Acknowledgements

E. Veldeman and F. Vanhaecke are indebted to the Nationaal Fonds voor Wetenschappelijk Onderzoek (N.F.W.O.) for financial support.

5. REFERENCES

1. E.N. Pentcheva, An. Idrol., 1967, 5, 90.

2. D. Beauchemin and S.S. Berman, Anal. Chem., 1989, 61, 1857.

3. D. Beauchemin, J.W. McLaren, A.P. Mykytiuk and S.S. Berman, Anal. Chem., 1987, 59, 778.

4. D. Beauchemin, J.W. McLaren, A.P. Mykytiuk and S.S. Berman, J.A.A.S., 1988, 3, 305.

5. J.R. Garbarino and H.E. Taylor, Anal. Chem., 1987, 59, 1568.

6. G.E.M. Hall, C.W. Jefferson and F.A. Michel, J. Geochem. Expl., 1988, 30, 63.

7. J.M. Henshaw, E.M. Heithmar and T.A. Hinners, Anal. Chem., 1989, 61, 335.

8. B. Sansoni, W. Brunner, G. Wolff, H. Ruppert and R. Dittrich, Fres. Z. Anal. Chem., 1988, 331, 154.

9. W.C. Blommaert, Trace element geochemistry in geothermal waters and in waters related to ore deposits, 1983, Ph.D. dissert., University of Antwerp (U.I.A.), Belgium.

10. C. Vandecasteele, M. Nagels, H. Vanhoe and R. Dams, Anal. Chim. Acta, 1988, 211, 91.

11. S.H. Tan and G. Horlick, Ap. Spec., 1986, 40, 445.

12. J.C. Miller and J.N. Miller, Statistics for Analytical Chemistry, 2nd ed. Wiley & Sons, New York, 1988.

13. J.D. Hem, Study and Interpretation of the Chemical Characteristics of Natural Water, 3th ed. U.S. Geol. Surv. Water-Supply Paper 2254, 1985, p. 263.

Semi-quantitative Estimation of Some Elements in Standards and Drinking Water by ICP-MS

Erland Johansson and Torsten Liljefors

DEPARTMENT OF RADIATION SCIENCES, DIVISION OF PHYSICAL BIOLOGY, BOX 535, S-751 21 UPPSALA, SWEDEN

1 INTRODUCTION

Estimation of elements by ICP-MS can be made in semiquantitative and quantitative modes. The latter include, multielement calibration, standard addition, and isotope dilution. According to VG Elemental, the semi-quantitative mode may yield results with a relative error of 30% or less. Work by different authors[1,2] suggested that better results could be achieved. After correcting the Saha-factors in the current data base by a standard solution, it may be possible to calibrate for individual differences of the ICP-MS instrument and we think that still better results can be achieved. To evaluate this possibility we examined some reference materials by the semi-quantitative mode.

2 MATERIAL AND METHODS

Of the reference material 300 mg were weighed accurately, then digested over night at 9°C in a mixture of 2.5 ml nitric acid (Merck, Suprapur) and 2.5 ml hydrogen peroxide (30%, Merck Suprapur) in closed teflon bombs. The residues were digested in a microvawe oven then chilled and the clear solution transferred to a volumetric flask and spiked with In to give 100 ppb of In and 3% nitric acid. All the calibration elements were obtained from Johnson & Matthey (Specpure).
The blank consisted of 100 ppb In and 3% nitric acid. Reference water 1643b was spiked with In to give 100 ppb. The drinking water was spiked with 100 ppb In and nitric acid diluted to 3% was added.

Instrumental set up for ICP-MS, VG PQ 1.
Forward power: 1400 W, reflected power: < 10 W.
Sample cone (Ni): 1.0 mm; skimmer cone (Ni) 0.75 mm.
Spray chamber: 10°C. Ar gas: nebuliser: 0.7 ml/min,
coolant gas: 13 ml/min, auxiliary gas 0.5 ml/min,
sample flow 1.0 ml/min, sampling distance: 10 mm from load coil.
Vaccuum controls: expansion stage 1.8 mbar, intermediate < 10^{-4} mbar, analyser: 5×10^{-6} mbar. Data acquisition: scanning mode (200 sweeps), run time 66 s.

3 RESULTS AND DISCUSSION

We used a mixture containing 100 ppb of Be, Mg, Co, Rb, Mo, In, La, Ho, Pb and U as reference elements for the mass response calibration curve.
The calibration solution included also B, Na, Al, Ca, Ti, V, Cr, Mn, Fe, Cu, Zn, As, Se, Zr, Cd, Ba, Ce, Hg and Bi. The mass response calibration curve of Be, Mg, Co, Rb, Mo, In, La, Ho, Pb and U was checked and if the points fitted the curve, the concentrations of the elements were estimated (VG, issue 3.1A).
The mass response calibration curve may vary in shape as shown in Figures 1, 2, 3.

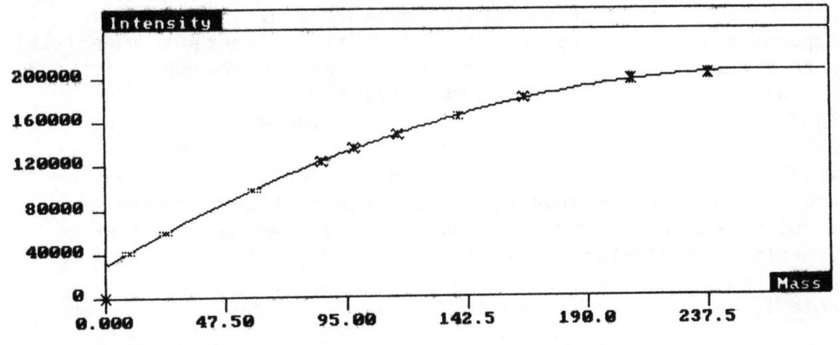

Figure 1 Mass response calibration curve 1

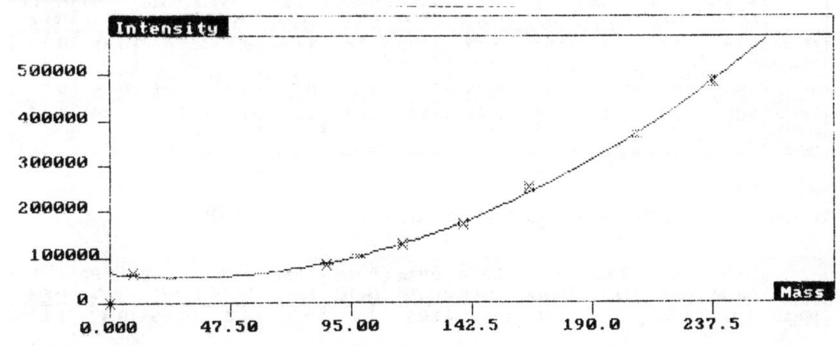

Figure 2 Mass response calibration curve 2

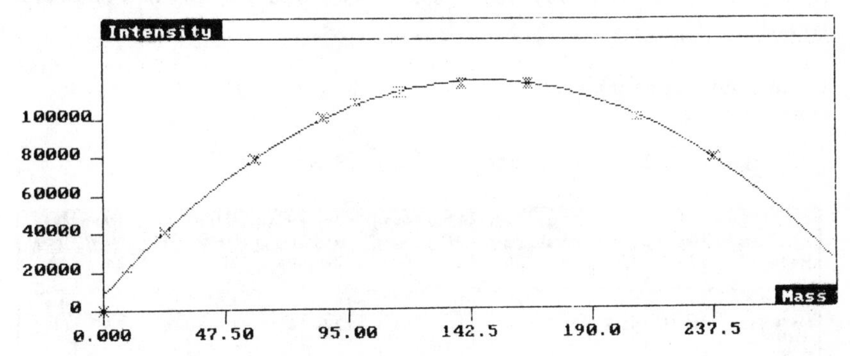

Figure 3 Mass response calibration curve 3

All runs were made on the same concentrations of the elements, same scanning parameters, flow rate, and forward power but at different times. The different shapes of the mass response calibration curves may have several explanations. If the temperature in the plasma varies then the ionisation of the beam ions may also vary. VG PQ 1 has a moveable torch box which may contribute to the variations. Even if the position is carefully centered minor differences may occur and deviate the ion beam off centre. The presence of inorganic salts in the plasma may also influence the shape of the curves but in these tests, all elements were of the same concentration. Differences in tuning conditions and choice of internal standard may also contribute to the shape of the mass response curve. After calibration, all the mass response curves gave satisfactory results on reference materials. To test the validity of the estimations by the semi-quantitative mode after a Saha calibration we estimated the elements in standard reference materials. Different standards including bovine liver (1577a), oyster tissue (1566) and reference water (1643b) were used for the evaluation. In general, the results agreed well with the labelled contents except in the case of selenium. Table 1 shows the results of the analysis of some elements in bovine liver, 1577a.

Table 1 Trace elements in bovine liver, SRM 1577a

	^{24}Mg (µg/g)	^{44}Ca (µg/g)	^{55}Mn (µg/g)	^{57}Fe (µg/g)	^{59}Co (µg/g)	^{63}Cu (µg/g)	^{66}Zn (µg/g)	^{85}Rb (µg/g)	^{88}Sr (µg/g)	^{98}Mo (µg/g)	^{111}Cd (µg/g)	^{208}Pb (µg/g)
Mean (n=5)	549	116	9.7	165	0.24	145	109	11.8	0.132	3.6	0.45	0.17
S	58	19	0.4	12	0.04	17	8	0.7	0.02	0.08	0.06	0.05
1577a, certif. value	600	120	9.9	194	0.21	158	123	12.5	0.138	3.5	0.44	0.135

Table 2 Trace elements in oyster tissue, SRM 1566

	^{51}V (µg/g)	^{52}Cr (µg/g)	^{55}Mn (µg/g)	^{57}Fe (µg/g)	^{63}Cu (µg/g)	^{66}Zn (µg/g)	^{85}Rb (µg/g)	^{88}Sr (µg/g)	^{111}Cd (µg/g)	^{202}Hg (µg/g)	^{208}Pb (µg/g)	^{238}U (µg/g)
Mean (n=5)	2.1	0.52	16.4	170	56	805	3.9	9.3	3.2	0.054	0.42	0.094
S	0.5	0.1	0.7	16	5	56	0.3	0.4	0.1	0.02	0.03	0.007
1566, certif. value	2.3	0.69	17.5	195	63	852	4.45	10.36	3.5	0.057	0.48	0.116

Table 2 presents the analysis of elements in oyster tissue, reference material 1566.

Table 3 Trace elements in water, SPM 1643b

	^{9}Be (ng/g)	^{11}B (ng/g)	^{51}V (ng/g)	^{55}Mn (ng/g)	^{59}Co (ng/g)	^{63}Cu (ng/g)	^{66}Zn (ng/g)	^{88}Sr (ng/g)	^{98}Mo (ng/g)	^{111}Cd (ng/g)	^{138}Ba (ng/g)	^{208}Pb (ng/g)	^{209}Bi (ng/g)
Mean (n=5)	19	93	33	38	28	27	67	222	96	20	39	24	10
S	2	9	5.6	13	2	7	5	16	2	1.5	2	2	0.4
1643b, certif. value	19	(94)	45.2	28	26	21.9	66	227	85	20	44	23.7	(11)

Table 3 gives the concentration of Be, B, Cr, Mn, Cu, Zn, Co, Sr, Mo, Cd, Ba, Pb in Reference water 1643b.
The difference between the certified value of SRM:s, 1577a, 1566 and 1643b and that found by ICP-MS is relatively small, relative error is about ± 10%.

Table 4 Trace elements in drinking water

	^{9}Be (ng/g)	^{11}B (ng/g)	^{51}V (ng/g)	^{55}Mn (ng/g)	^{59}Co (ng/g)	^{63}Cu (ng/g)	^{66}Zn (ng/g)	^{88}Sr (ng/g)	^{98}Mo (ng/g)	^{111}Cd (ng/g)	^{138}Ba (ng/g)	^{208}Pb (ng/g)	^{209}Bi (ng/g)	^{238}U (ng/g)
I	-	16	<1	5	<1	36	47	46	<1	<1	14	<1	-	<1
II	-	52	<1	390	<1	7	21	186	<1	<1	22	<1	-	<1
III	-	59	<1	256	<1	16	22	129	3	<1	10	<1	-	1.4
IV	-	-	<1	4	<1	20	16	91	7	<1	54	<1	-	27
V	-	-	<1	67	<1	34	360	121	2	<1	44	<1	-	14
VI	-	45	<1	6	<1	21	4	164	5	<1	37	<1	-	26
VII	-	24	1	19	<1	83	18	206	5	<1	52	<1	-	33
VIII	-	20	2	-	-	32	8	<1	5	<1	<1	<1	-	29
IX	-	3	<1	2	<1	2	20	67	1	<1	8	<1	-	<1

I. COUNTRY WELLS, WEST COAST
II. " "
III. " "
IV. " ", EAST COAST
V. " ",
VI. TAP WATER, UPPSALA
VII. " " , "
VIII." " , ", SOFTENED
IX. TAP WATER, STOCKHOLM

Table 4 shows the concentration of Be, B, V, Mn, Cu, Zn, Co, Sr, Mo, Cd, Ba, Pb in drinking water of some households as estimated by using corrected Saha factors.
In Table 4 the elements which are not detectable are marked -. Interestingly we found about 30 ppb U in the drinking water of Uppsala. This drinking water passes through boulder ridges which may be one explanation for the increased uranium concentration. Using the semi-quantative mode may reveal the presence of elements that not generally investigated; knowledge of their presence however might be of protective value.
The name semi-quantitative analysis is misleading since after proper mass response calibration, quantitative results can be obtained after testing standard reference materials (SRM). A better name would be conditional Saha estimation since quantitative results are achieved. The use of conditional Saha estimation may be useful for the screening analysis of drinking water with low content of organic material.

The authors acknowledge with gratitude the donation of the Crafoord foundation which made this study possible. One of the authors is indebted for financial support to Astra Läkemedel AB, Södertälje. The authors are indebted to W.G.P. Mair for criticism of the manuscript.

REFERENCES

1. A.L. Gray, R.S. Houk and J.G. Williams, J. Anal. At. Spectrom., 1987, 2, 13.
2. T.D.B. Lyon, G.S. Fell, R.C. Hutton and A.N. Eaton, J. Anal. At. Spectrom., 1988, 3, 265.

ICP-MS Analysis for Weakly Bound Gold in Humus Samples: an Aid to Gold Exploration in Areas of Glacially Transported Overburden

B.J. Perry and J.C. Van Loon

DEPARTMENT OF GEOLOGY, UNIVERSITY OF TORONTO, CANADA M5S 3B1

1 ABSTRACT

Humus developed on glacially transported overburden can contain both exotic gold (strongly bound) and indigenous gold (weakly bound). Total determinative methods for analyzing exploration humus samples for gold, such as neutron activation analysis of irradiated compressed humus briquettes, cannot discriminate between these two fractions. The concentration of weakly bound gold, as extracted from exploration humus samples by 1.2M HCl, increases with proximity to bedrock gold mineralization at both study sites. Since the concentrations of weakly bound gold extracted from exploration humus samples are less than one ppb, cost effective analysis can only be obtained through ICP-MS.

2 INTRODUCTION

The utility of partial dissolution techniques in geochemical exploration for base metals has long been recognized. Chao provides an excellent review of these techniques [1]. In the case of gold, however, there is little guidance in the literature regarding either partial dissolution techniques or their respective exploration applications. A notable exception is the work by Roslyakov which indicates that gold associated with organic matter (humus) in proximity to gold mineralization may be associated with humin, humic, and fulvic acid components [2]. Away from mineral occurrences gold in humus appears to be associated only with the fulvic acid component. Gregoire describes a technique for extracting gold bound in organic matter using sodium hypochlorite digestion, but does not pursue the relationship, if any, between the concentration of gold in this fraction and proximity to mineralization [3]. Early work by Curtin *et al* indicated that very small concentrations of gold were water leachable from mull humus in the Empire District of Colorado [4].

If the development of a partial dissolution technique for gold in humus is to lead to practical exploration application, it must yield a relatively simple, cost competitive technique that reliably and accurately targets overburden covered subcropping orebodies. Exploration in the extensive glacial debris covered areas of Canada would particularly benefit from a refinement in the humus geochemical survey method as it is now routinely applied. Current practise is to determine only total gold concentration. Most often this is accomplished by instrumental neutron activation analysis of irradiated compressed humus briquettes, or by mixed mineral acid digestion of humus (or ashed humus) followed by graphite furnace atomic absorption spectroscopy to determine the gold concentrations in the resultant solutions. However, it must be recognized that in transported overburden exploration terrains there are ultimately two gold fractions present in the humus horizon: gold that is derived from underlying bedrock gold mineralization and then carried to the humus horizon by biological mechanisms, i.e. the "indigenous" gold fraction; and gold that is transported to the site along with the overburden, bears no relation to the underlying bedrock and is therefore termed the "exotic" gold fraction. It is impossible, of course, for total gold determinative methods to distinguish between these two fractions. Most often, the erratic distribution of the exotic gold fraction masks the much weaker biogeochemical indigenous gold signature of underlying gold mineralization. Therefore, the development of a partial extraction that obtains the biologically emplaced indigenous gold component, while largely ignoring mechanically emplaced exotic gold component, would be an important improvement that would have wide application to gold exploration since much of the favorable gold exploration terrain in Canada, and elsewhere, contains humus horizons developed on glacially transported overburden.

3 EXPERIMENTAL

Preliminary studies

Early experimentation by the authors was entirely empirical and consisted of testing various simple to moderately complex extractions for indications of a positive relationship between gold concentration in the extracts and proximity to gold mineralization. A positive relationship was observed for 1.2M HCl extractable gold concentrations in the outcrop scale preliminary study at the Aubrey Twp. study site. Weak HCl extractable gold concentrations increased with increasing proximity to the gold mineralization (80 ft, 0.7ppb; 20 ft, 1.5ppb and 2.0ppb; 6ft, 4.0ppb). This encouragement prompted expanding the scale of the study at the original property, and also encouraged us to analyze humus samples from another study site where known bedrock gold mineralization is also covered by glacially transported overburden.

Study Site I: Aubrey Township, Ontario, Canada

Gold bearing quartz veinlets are exposed in a small outcrop of metabasalt (chlorite schist) located near Dryden, Ontario. The veinlets occupy a zone of fracturing approximately 3 m wide that yielded 0.17 oz. Au/ton across 2m in drill core (Voyager Explorations Ltd.). The humus is thinly developed (max 3 cm) on very fine, clean, sandy glacially transported overburden, likely glaciolacustrine in origin. Vegetation contributing to the humus is mixed coniferous forest, mostly pine and spruce with a minor deciduous presence. Alder, mosses, ferns and grasses comprise the typical underbrush. The sandy till provides good drainage at all the humus sampling sites. Five test pits show only fine sand overburden to the maximum depth reach (14') of the backhoe employed. Nineteen humus samples were collected from 13 sampling sites. Triplicate humus samples were collected at three sites.

Study site II: Van Horne Twp., Ontario, Canada

The Van Horne Township study site is 4 miles east of the Aubrey Twp. study site. Here, humus is also developed on glaciolacustrine sands, but more outcrop is present than at the Aubrey Twp study site. The humus layer is generally less than 3cm. The vegetation differs only slightly from the Aubrey site in that there is more birch present, and less poplar. Subcropping gold mineralization is known to exist at site 2 (drill intersection at bedrock/overburden interface 0.05 oz Au / ton across 3.0', overburden depth = 18', Van Horne Gold Exploration Ltd.). Thirty-four humus samples were collected from 21 sites. Triplicate humus samples were collected at five sites.

Humus sample collection and preparation

At each study site, a control line was cut through the underbrush across the strike of the known bedrock gold mineralization. Humus, identified as black soil containing less than 20% partly decomposed relict vegetation (needles, twigs), was collected by hand into paper bags at sites spaced approximately fifty feet apart along the line crossing the overburden covered gold mineralization. Approximately 100 grams (dry weight) was collected per sample site. On each property at several sites well removed from the gold mineralization, three additional humus samples were collected at each of these sampling sites. These triplicate samples are used to calculate the expected average local natural variation, which in turn is used to calculate the threshold above which gold concentrations are considered anomalous. The samples are dried in an oven at 70° C. As much as possible of the dried sample is passed through an 80 mesh sieve. The -80 mesh fraction is analyzed.

Equipment

During the course of the study, two ICP-MS systems were used to determine the natural isotope of gold, ^{197}Au. At the Department of Geology, University of Toronto, Canada, a quadrupole mass filter based instrument (Sciex ELAN 250, prototype) was used. Details of this system have previously been published [5]. The ion source was an inductively coupled plasma generated using a Plasma Therm Model ICP 2500 power supply, impedance matching network, and torch. The incident power was 1.25 or 1.50 kW. The reflected power was maintained as close to zero as possible. Mass selection was accomplished using a Tracor Northern Model 1251 Ramp Generator/ Conditioner synchronized with a Tracor Northern Model TN 7200 Multichannel Analyzer with 4096 channel data memory. Signal detection was performed with a pulse counting CEM and a pulse shaping and discriminator preamplifier. The dynamic range was 1 to 10^6 counts per second. The quadrupole was operated at a resolution to give peak widths of approximately 1 amu at 10% peak height. Additional analyses, and cross-check analyses, were performed using a commercial version Sciex ELAN 250, located at the Ontario Geological Survey Geoscience Research Laboratory, Toronto, Canada (primary operator, W. Doherty).

Au Standards

A stock 1000ppm gold standard was prepared by dissolving 0.100 metallic gold in 4ml of warm AR in a 100ml volumetric flask. After the gold had dissolved, the flask was brought up to volume with 1.1M HNO_3. This stock standard was stored in a brown plastic bottle under constant refrigeration (4°C). All subsequent gold standards were prepared from this stock standard, or dilutions thereof, and were carried in 1.1M HNO_3. Intermediate 10ppm and 1ppm stock solutions were made. Low concentration standards (2.5ppb, 1.0ppb, 500ppt, 200ppt, 100ppt) were made daily by serial dilution of the intermediate stock solutions. All standards were kept under refrigeration (4°C) when not in use.

Procedures

Aqua regia (AR) extraction for strongly bound gold. Three grams of -80 mesh humus sample are weighed into a 250 ml pyrex beaker. Forty ml of AR, prepared with de-ionized distilled water (ddw), ultrapure HCl and ultrapure HNO_3, are added to the beaker. The beaker and its contents are agitated by hand at room temperature until reaction ceases (5-15 minutes). The beaker is placed on a hot plate at low temperature, and hand agitated until reaction ceases (10-20 minutes). The temperature is raised to low-medium heat and the beaker is hand agitated until reaction ceases (10-20 minutes). In case of excessive rapid frothing, a small volume of ddw is injected around the inside walls of the beaker from a wash bottle equipped with a very small diameter tip, and the beaker is temporarily removed from the heat. After this reaction has ceased the temperature is raised to medium heat and the beaker is hand agitated until the reaction again ceases. The beaker is left on the hot plate at medium temperature until only a few millilitres of solution remain. The beaker is transferred to a hot plate kept at very low heat and the contents are evaporated to near dryness, i. e. still damp. Caution must be exercised at this point not to allow the contents of the beaker to evaporate to dryness, since a portion of the gold chloride may volatilize under the high heat obtained after the temperature stabilizing effect of evaporation ceases. Equally detrimental is that baking conditions may drive off chlorine. Chlorine as chloride must be present in excess in order to facilitate the presence of gold as a soluble gold chloride. Thermal dissociation of the gold chloride complexes is also a possibility at dryness during inadvertent baking. Any contents that were baked cannot be used in the context of meaningful comparison between solutions, and should be discarded. The beakers are covered with parafilm and stored in a cool dark cupboard until the day that the determination will be made is at hand. The procedure is resumed by adding 10 ml of 1.2M HNO_3 to the beaker, and heating the contents on low heat for 5-10 minutes. The contents are then stirred with a glass rod, the lumps broken up and the whole of the bottom one third of the beaker scraped with a teflon policeman. The contents are filtered through a 9.0 cm diameter #40 Whatman filter paper into a 25 ml volumetric flask. The beaker is rinsed twice with 5ml 1.2M HNO_3. The rinses are passed through the filter into the flask. The filter is rinsed twice into the flask with 2 ml 1.2M HNO_3 each rinse. The volume in the flask is adjusted with 1.2M HNO_3. The solution is stored under refrigeration (4^oC). The gold determination should be made as soon as possible the same day by ICP-MS analysis. The gold concentration in the resultant solution is 8.33 times less than that in the starting material.

1.2M HCl extraction for weakly bound gold. Each 3g subsample of -80 mesh humus is shaken for thirty seconds in a test tube containing 10 ml 1.2M HCl (ultrapure HCl, ddw). The test tubes are allowed to stand in racks for one hour, during which time they are individually shaken for 10 seconds every 20 minutes. The contents of each is then filtered through #54 Whatman filter paper into a 100ml pyrex beaker. The test tube is rinsed twice with 5 ml (ddw), and these rinses are passed through the filter paper into the beaker. The filter paper and its contents are rinsed twice with 5 ml ddw, and these rinses are passed through the filter paper into the beaker. The solution is evaporated to near dryness over low heat of a hot plate. The beaker and its contents can be stored at this stage by covering with parafilm and keeping out of the light in a closed cupboard. The procedure is resumed by adding 3 ml HCl and 1 ml HNO_3 to each beaker. The solution is heated over low heat for fifteen minutes and evaporated to near dryness. The procedure may be interrupted at this point by covering the beakers with parafilm and storing in a cool, dark cupboard. It is strongly advised not to continue putting the sample to the final solution unless ICP-MS analysis can be accomplished within a few hours. The procedure is resumed by adding 5 ml of 1.1M HNO_3 to the beaker. The contents of the beaker are heated over low heat for five minutes, during which time they are occasionally agitated by hand. The solution is filtered through #40 Whatman filter paper into a 10 ml volumetric flask. The beaker is rinsed twice with 1.1M HNO_3. The rinse solutions are passed through the filter into the flask. The filter paper is rinsed twice with 1.1M HNO_3. The flask is brought up to volume with 1.1M HNO_3. The gold concentrations in the resultant solutions are determined by ICP-MS analysis.

Gold concentration determination by ICP-MS. The concentrations of 1.2M HCl extractable gold in the final solution is 3.33 times less than the gold concentration in the starting material. The gold concentrations in these solutions are present at ultra-trace concentrations (less than 1ppb). Therefore, cost effective analysis can only be obtained by direct analysis of the solution by ICP-MS. Other methods (INAA, GF-AAS) would require costly and/or error introducing gold preconcentration techniques to bring the level of gold concentrations up to measurable levels. *ICP-MS (University of Toronto):* Solutions are delivered directly from the test tube, after shaking, by inserting the nebulizer intake tube directly into the test tube to about one inch from the bottom of the test tube. Sample is delivered continuously for 30 seconds, after which the nebulizer intake is put into ddw. After the counts per second fall back to background level the next sample solution is aspirated. Standards are determined at the beginning and end of each run of approximately 25 sample solutions. The gold concentration is obtained by direct comparison to standard solutions. *ICP-MS (Ontario Geological Survey)*: Solutions are delivered either manually or by automatic sampler. Concentration values are obtained by the adjunct software using direct comparison with standard solutions.

Determination of Loss On Ignition (LOI)[6]. LOI was determined for each site sample using a 1g subsample of the dried -80 mesh humus material. The 1g subsample is ignited in a muffle furnace pre-heated to 150^0 C. The temperature is gradually raised to 600^0C and maintained for one hour. The sample is cooled in a desiccator and reweighed. The LOI is taken to be the difference in weight between the original sample and the weight of the ash. The ash is discarded.

Normalization of gold concentrations to 100% ashable organic concentration. The gold concentration in the humus sample is normalized to 100% ashable organic concentration by dividing the gold concentration obtained for the -80 mesh humus sample by the decimal fraction LOI obtained for a 1g subsample of the same -80 mesh exploration humus sample.

4 RESULTS

Study site I: Aubrey Township, Ontario

Threshold in this study takes into consideration the encounterable natural variation at an average site. Threshold is the mean sample value of all samples collected along the sampled line plus 2.5 times the within-site coefficient of variation of an average site multiplied by the sampled line mean, i.e.

threshold=line mean+2.5(mean within-siteCv)(line mean) (4-1)

The calculation of the threshold for the total (AR extractable) gold method uses the average within-site coefficient of variation as determined from data obtained from triplicate humus samples collected at sampling sites located more than 500 feet away from the known gold mineralization. The threshold was calculated to be 9.5 ppb. The line profile of the AR extractable gold distribution (Fig. 1) shows that only one site (2S, 16.2ppb)) is identified as anomalous. It is not the site nearest to the known gold mineralization. The total gold concentration for the site nearest to the gold mineralization (site 1S, 2.4ppb) is relatively low. In terms of accurately targeting the known gold mineralization, the total gold method was unsatisfactory.

Similarly, the calculation of the threshold for the 1.2M HCl extractable gold distribution also takes into consideration the expected natural variation at an average site. The average of the with-in site coefficients of variation of the 1.2M HCl extractable gold concentrations for the triplicate humus samples is used. The threshold for the 1.2M HCl extractable gold concentrations is calculated to be 0.88 ppb.

The line profile of 1.2M HCl extractable gold distribution (Fig. 2) shows three values that are high, but not anomalous since none exceeds the calculated threshold of 0.88 ppb Au . Two of the highest concentrations straddle the known gold mineralization. Site 1S is nearly coincident with the known gold

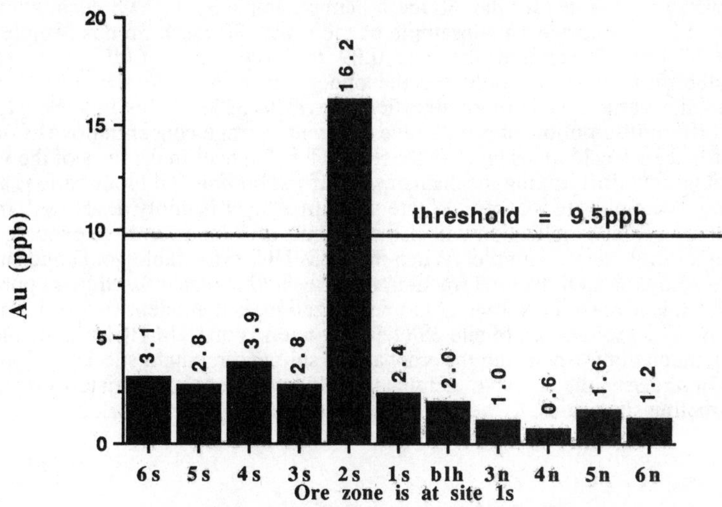

Fig. 1 Total Au (AR extractable), Aubrey Twp.

mineralization, while site 2S is 50 feet south of the gold mineralization. This technique accurately highlights the known gold mineralization but falls short of identifying the 1S site, or any site, as distinctly anomalous. Therefore, in terms of actually targeting the known gold mineralization, this treatment does not adequately emphasize the significance of site 1S.

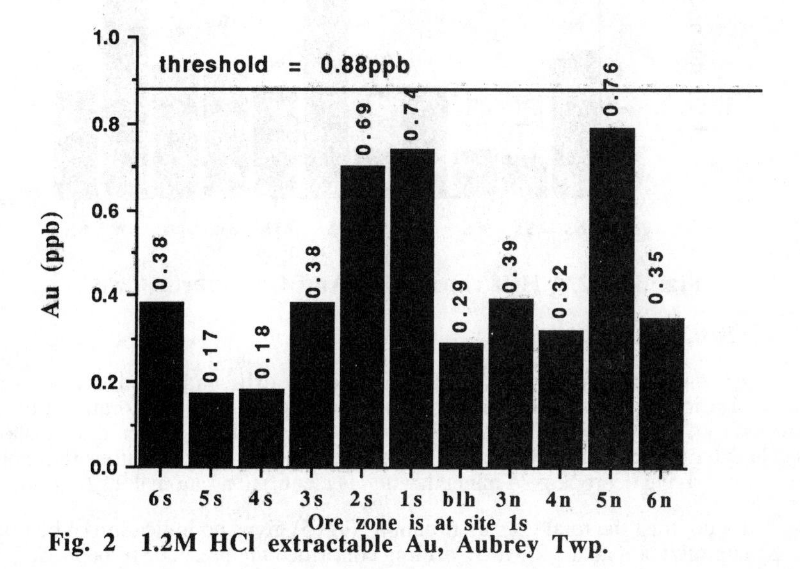

Fig. 2 1.2M HCl extractable Au, Aubrey Twp.

Improvement in the gold mineralization targeting ability of the 1.2M HCl extraction technique was gained by normalizing these gold concentrations to 100% organic matter content. This is done by dividing the 1.2M HCl extractable gold concentration obtained for the -80 mesh humus sample by the LOI decimal fraction obtained for a separate 1g subsample of the same -80 mesh humus sample. The range of LOI determinations was 0.09 to 0.64 g/g . LOI within-sample reproducibility was excellent for the one gram subsample size (avgCv=4%). Within-site variability is more significant (avgCv=31%). This is to be expected since different sampling sites will have different average concentrations of organic matter in their local humus layer as the result of differences in the rates of the various natural humus-drift mixing mechanisms (e.g. frost boiling and bioturbation) at each individual sample site locality. Where the humus layer is thinly developed, manual error during sample collection will also contribute drift (and therefore exotic gold) to the exploration humus sample. When the 1.2M HCl extractable gold concentration is divided by the LOI decimal fraction, the threshold is recalculated to 4.9ppb. Site 1S, which is nearest the known gold mineralization, is then identified as anomalous (Fig. 3). The prominence of site 2S (1.1ppb) in terms of 1.2M HCl extractable gold is very much diminished, and the emphasis is shifted directly to site 1S (7.9ppb). In terms of targeting the known mineralization, this treatment appropriately emphasizes the sampling site nearest to the known ore by indicating it as anomalous.

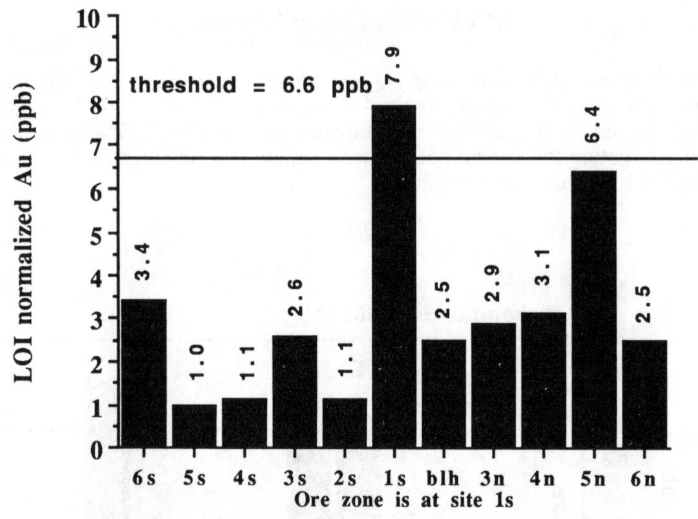

Fig. 3 1.2M HCl extractable Au/LOI, Aubrey Twp.

Study site II: Van Horne Twp.

At the other study site four miles east, drill indicated minor gold mineralization (0.05 oz Au/ton across 3 feet, Van Horne Gold Exploration Ltd.) is covered by 18 feet of very fine glaciolacustrine sands at site 2. When the threshold is calculated for LOI adjusted 1.2M HCl extractable gold concentrations (8.5 ppb), the site overlying bedrock gold mineralization is identified as anomalous (Fig. 4).

By contrast the total gold distribution (Fig. 5) gives no indication of bedrock gold mineralization at site 2, where drilling confirmed the presence of bedrock gold mineralization. Rather, the total gold distribution indicates site 15 as anomalous,

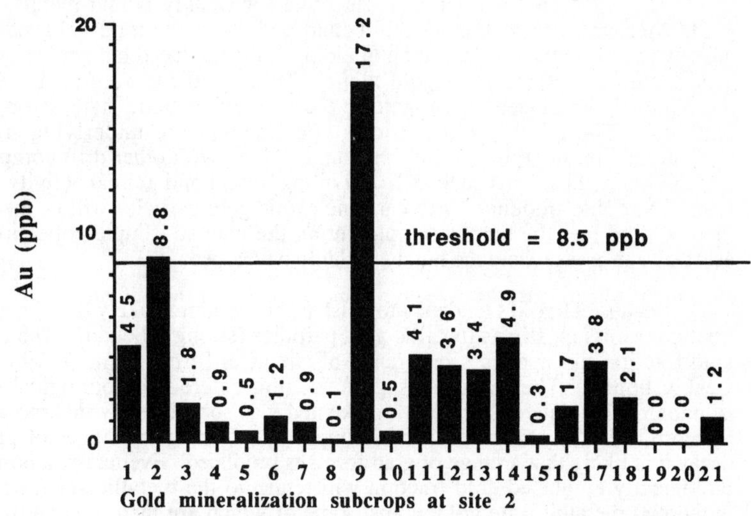

Fig. 4 1.2M HCl extractable Au / LOI, Van Horne Twp.

although there is no known gold mineralization at or near this site (Van Horne Gold Exploration Ltd.).

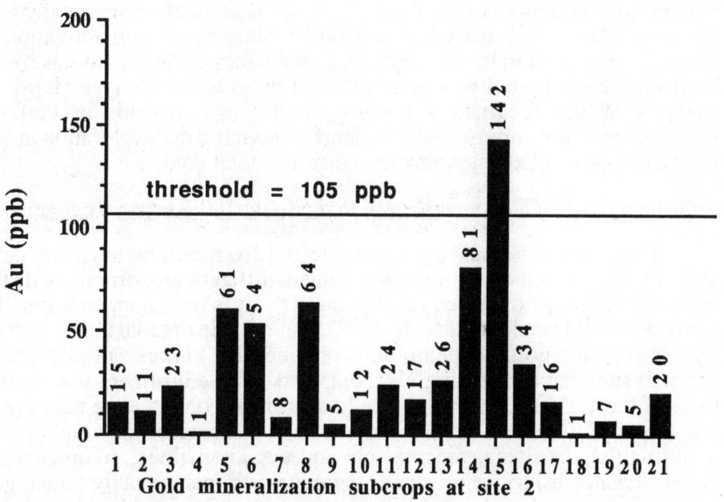

Fig. 5 Total Au (AR extractable), Van Horne Twp.

5 DISCUSSION

Extraction of weakly bound gold

The 1.2M HCl extraction is selective for weakly bound metals, including gold. Measurement of this weakly bound portion of the total gold may be more directly related to the concentration of biologically emplaced indigenous gold than is a measurement of the total gold alone. Total gold measurement will include mechanically emplaced exotic gold in the form of metallic state, strongly bound particles. The exotic gold particles contained in the underlying transported overburden are incorporated into the humus along with other drift components by soil mixing mechanisms such as frost boiling, insect and animal activity, wind and rain. A variable amount of drift carrying exotic gold particles will be inadvertently incorporated into the humus sample during the manual sample collection routine, especially in areas where the humus is thinly developed.

Because of gold's tendency to exist as the metal it is likely that the exotic gold fraction occurs mostly as metallic gold particles (strongly bound). The indigenous gold fraction likely occurs in a range of chemical forms, some of which may be weakly bound. The indigenous gold fraction entered the biological system in solution during aqueous nutrient uptake and was subsequently incorporated, atom by atom, into the structure of the living plant. During the stages of progressive decay it is likely that a range of gold forms is produced. Eventually a portion of the biologically emplaced gold fraction will return to the metallic state, occurring as individual metallic state gold atoms, some of which are then likely adsorbed onto clay particles, carbonaceous material and other suitable particles within the humus layer. These biologically emplaced gold atoms may also become weakly bound within various substances in the humus layer, e.g.Fe-Mn hydroxides.

Both the weakly bound gold forms and the adsorbed gold atoms are amenable to extraction by weak HCl, while strongly bound gold forms (exotic gold particles) are not. The difference in chemical form between significant portions of each gold fraction (indigenous vs exotic) is the basis for analytical distinction between the two fractions. Total gold analytical techniques, such as neutron activation analysis of irradiated compressed humus briquettes, or decomposition of humus (or humus ash) by strong mineral acid mixtures followed by graphite furnace atomic absorption analysis of the resultant solutions, cannot provide this crucial distinction. Unfortunately, the commercial laboratories servicing the exploration industry do not offer any type of gold determination other than total gold.

Adjusting 1.2M HCl extractable Au to represent 100% organic content.

Exploration humus samples collected from humus horizons developed on glacially transported overburden will contain different proportions of drift (inorganic matter) and humus (decayed organic matter). It is necessary to normalize the HCl extractable gold concentrations to 100% total ashable organic matter before the 1.2M HCl extractable gold concentrations obtained for humus samples collected from different sites can be meaningfully compared. This adjustment sharply increases the focus of the 1.2M HCl extractable gold distribution onto the ore zones (e.g. compare Fig. 2 with Fig. 3), and equally as important, it increases the emphasis sufficiently to distinguish the sites nearest the ore zones as anomalous. In other words, at each of the sites nearest to the ore zones there is much more weakly bound gold per unit of organic matter than there is for several hundred feet in either direction away from the ore zones. This indicates that biogeochemical communication exists between the bedrock gold mineralization and the humus layer even though they are separated by a blanket of glacially transported overburden upto 18' thick.

6 CONCLUSIONS

The routine total-gold-only humus survey used in explorations for bedrock gold mineralization in areas of glacially transported overburden should be superseded by surveys using analytical procedures that provide more discriminatory information than can be obtained by total-gold-only analysis. Analytical discrimination must be provided between the biologically emplaced indigenous gold fraction, which is useful for locating subcropping gold deposits that are in biogeochemical communication with the humus layer, and the mechanically emplaced exotic gold fraction, which is not only not useful, but often provides misleading information. It appears that the 1.2M HCl extraction technique takes into solution a portion of the indigenous gold fraction of exploration humus samples, while largely ignoring the exotic particulate gold. Thus, this new method is able to take advantage of the small accumulations of biologically emplaced indigenous gold in exploration humus samples that build up in the humus layer directly over bedrock gold mineralization as a result of biogeochemical communication between a subcropping gold deposit covered by foreign overburden and the humus horizon directly above it. Accuracy and contrast are improved by adjusting the 1.2M HCl extractable gold concentrations obtained from exploration humus samples to reflect 100% ashable organic content. This is done by dividing the 1.2M HCl extractable gold concentration obtained for each humus sample by the LOI decimal fraction obtained from a 1g subsample of the same humus sample. Per unit of organic matter, anomalous proportions of 1.2M HCl extractable gold exist in the humus horizon directly above the known bedrock gold mineralization at each study site.

ACKNOWLEDGEMENTS

Thanks are due Perkin-Elmer SCIEX and Professor J. B. French for the use of the ELAN 250 (prototype) ICP-MS located at the, Department of Geology, Earth Sciences Center, University of Toronto. We are most grateful to the Ontario Geological Survey for determinations made using their SCIEX 250 ICP-MS under the friendly direction and expert guidance of Mr. W. Doherty. Dr. C. Park assisted in the supervision of ICP-MS determinations made at the Department of Geology, University of Toronto.

REFERENCES

1. Chao, T.T., Geochem. Explor., 1984, 20, 101-135.
2. Roslyakov, N.A., 1983, 10th Intl.Geochemical Symposium, Helsinki; pp 66-67.
3. Gregoire, D.C., J. Goechem. Explor., 1985, 23, 299-313.
4. Curtin, G.C., *et al*, U. S.Geol. Survey Circ 562, 1968.
5. Douglas, D., Quan, E. and Smith, R. G., Spectrochim Acta, 1983, 38B, 39.
6. Lye, J., M.Sc. thesis (unpublished), University of Toronto, 'The choice and application of selective extractions to elucidate the chemical forms of trace elements in some lake sediment samples', 1982.

A Procedure for the Determination of ^{99}Tc in Environmental Samples by ICP-MS

Ihsanullah and B.W. East

SCOTTISH UNIVERSITIES RESEARCH AND REACTOR CENTRE, EAST KILBRIDE, GLASGOW G75 0QU, UK

1 INTRODUCTION

The existence of a missing element with atomic number 43 was first predicted by Mendeleev in 1869 Initially, it was named "eka-manganese" from its position in periodic table of the elements. In 1937 the Italian physicists Perrier and Segre were the first to prepare the element by bombarding molybdenum with deutrons in the University of California cyclotron[2].

The element was the first to be artificially created by man and was given the name "technetium" meaning artificial[3]. The name technetium was officially confirmed at a convention of chemists held on September 2-5, 1949 in Amsterdam [4,5].

Technetium has no stable isotopes. Twenty-one radioactive isotopes, along with seven isomers are known. There are three long-lived isotopes, ^{97}Tc ($t_{½}$=2.6 x 10^6a), ^{98}Tc ($t_{½}$=4.2 x 10^6a)) and ^{99}Tc ($t_{½}$ = 2.1 x 10^5a). Of these radionuclides only ^{99}Tc is a fission product[6-7]. ^{99}Tc is the most important technetium isotope environmentally[8] and is a weak beta emitter with $E_{\beta}max$ = 0.292 MeV and specific activity 630 kBq mg^{-1}.

Sources of ^{99}Tc

^{99}Tc is the daughter product of ^{99}Mo, which is formed by thermal neutron fission of ^{239}Pu (fission yield 5.9%) and ^{235}U (fission yield 6.1%) and also by neutron capture in ^{98}Mo[9]. ^{99}Tc is also formed in high abundance form thermal neutron fission of ^{233}U (4.8%), and fast neutron fission of ^{239}Pu (5.9%), ^{238}U (6.3%) and ^{232}Th (2.7%)[10]. ^{99}Tc is produced from nuclear detonation tests, nuclear fuel plants, nuclear power plants and the use of ^{99m}Tc in nuclear medicine. The amount of ^{99}Tc in certain U ores can be somewhat higher than that expected from the spontaneous fission of ^{238}U alone. This is due to the production of ^{99}Tc by neutron-induced fission of ^{235}U[11]. To summarise, ^{99}Tc is produced both naturally and by man.

^{99}Tc in the Environment

Because of its long term radiological implications a knowledge of the behaviour of ^{99}Tc in the environment is important in the assessment of the impact of the nuclear industry. The difficulty of its analysis however data on technetium in the environment are hitherto limited.

Methods for the Measurement of ^{99}Tc

Various physical and chemical properties of ^{99}Tc such as ß-activity; complex formation; absorption bands of X-rays, UV and IR radiation; mass to charge ratio etc. can be used for the qualitative, as well as the quantitative analysis of technetium. Many papers and reports[4,5,7,9] have reviewed the available methods.

In general, the different techniques can be arranged in the following order of sensitivity:

Method		wt. of Tc measurable (g)
1.	X-Ray Fluorescence	10^{-5}
2.	Gravimetry/Classical	10^{-6}
3.	Infra-Red Spectrometry	10^{-6}
4.	Spectrophotometry	10^{-6} - 10^{-7}
5.	Emission Spectroscopy	10^{-8}
6.	Polarography	10^{-8} - 10^{-9}
7.	Atomic Absorption	10^{-8} - 10^{-9}
8.	Radiometry/ß-Counting	10^{-9}
9.	Mass Spectrometry	10^{-10} - 10^{-11}
10.	Resonance Multiplication Ionisation MS	10^{-11} - 10^{-12}
11.	Neutron Activation	10^{-11} - 10^{-12}
12.	Inductively Coupled Plasma - MS	10^{-12}

Clearly the selection of a method for the determination of Tc is mainly dependent on the concentration; the type and number of samples, their location, the availability and cost of facilities and the type of information required.

Inductively Coupled Plasma Mass Spectrometry (ICP-MS)

There are many methods in the literature for radiochemical analysis of technetium, but most of them are not designed to deal with the low levels of activity likely to be found in environmental samples. The work described in this paper has been directed towards the development of an accurate and precise technique for environmental samples, and with the advent of ICP-MS as a highly sensitive method of detection, this has been investigated and successfully applied to ^{99}Tc analysis.

2. EXPERIMENTAL

Method

Measurements were carried out on a VG Elemental ICP-mass spectrometer (Plasmaquad PQ1) installed at SURRC, East Kilbride, Glasgow.

Standards and Yield Tracers

In order to standardise the analysis method and to determine the chemical yield of the radiochemical procedures various standards and tracers were used in the determination of technetium.

Technetium-99: ^{99}Tc, as NH_4TcO_4 was obtained from Amersham International plc, and used to determine the radiochemical yield of separation and decontamination procedures.

Technetium-99m: The γ-emitting isotope ^{99m}Tc ($t_{½}$ = 6hrs) was also used in some stages for the optimisation of radiochemical yield. ^{99m}Tc was obtained from the Radiochemical Dispensary at the Western Infirmary, Glasgow.

Technetium-95m: Clearly ^{99m}Tc was not suitable for lengthy procedures because of its short half-life. In later work ^{95m}Tc ($t_{½}$=60d) was used instead to determine yield and to optimise procedures and as a spike when examining environmental samples. ^{95m}Tc was supplied by the Isotope Division, Harwell, UK.

Ruthenium: Atomic absorption (AA) grade standard Ru as $RuCl_3$ solution was supplied by Johnson Mathhey. Ru was used to determine Ru decontamination factors.

Rhodium: AA grad standard Rh was obtained from Johnson Mathhey, ^{103}Rh was used as an internal standard to check any drift in the response of the ICP-MS instrument.

THE DEVELOPMENT OF THE METHOD

The optimised method consisted of the following three main parts which are now described and for which detailed procedures are given.

A. SAMPLE TREATMENT

The Tc from 5-30l of filtered (0.2μm) sea water or filtered rain water was concentrated on Dowex 1-x8 anion exchange resin (5cm^3 slurry in column form), then Tc was eluted with 40-45 cm^3 12M HNO_3, and the eluate was evaporated and redissolved in 50-100cm^3 2M H_2SO_4, containing 2cm^3 of H_2O_2.

Between 1-30g dried (110°C) biota samples were leached with 9M HNO_3 (180-720cm^3) by refluxing. The leachate was evaporated on hot plate until brown fumes ceased to be evolved. The residual solution was filtered (Whatman No 1) and the filtrate was evaporated and redissolved in 50-100cm^3 2M H_2SO^4 containing 2cm^3 if H_2O_2. The soils and sediments (10-20g) were ashed at 600°C in a furnace overnight for 24 hours. The ash was transferred to a 250cm^3 and 50-100cm^3 of 2M H_2SO_7 containing 2cm^3 H_2O_2 added. The solution was filtered (whatman No1).

B. ELIMINATION OF ISOBARIC INTERFERENCES WITH ^{99}Tc

Since the ICP-MS method is based on the measurement of the mass of an element, isobaric interferences as mass 99 must be eliminated as far as possible. For ^{99}Tc, two isobaric nuclides (^{99}Mo & ^{99}Ru) are important. ^{99}Mo is a radioactive isotope ($t_{½}$=67hrs) and will have decayed after a few days. ^{99}Ru, however, is stable and is a key problem. It has a 12.72% natural abundance and is prevalent in the environment. Ru was removed by precipitation and solvent extraction.

C. THE PREPARATION OF SOLUTION FOR ICP-MS ANALYSIS

For ICP-MS analysis the final solution, has to be in a suitable form: < 2% in HNO_3; a salt concentration of <0.1% and colourless. After treatment to remove Ru, the resulting solution was adjusted to pH7 and passed through 5cm^3 slurry of resin (Dowex 1-X8). The column was washed thoroughly with cold and then hot distilled water to remove salts and Tc was stripped with 40-45cm^3 12M HNO_3. The eluate was evaporated, the residue was redissolved in 2% HNO_3 and filtered if necessary.

Details of radiochemical procedures

A. Initial sample treatment to isolate technetium

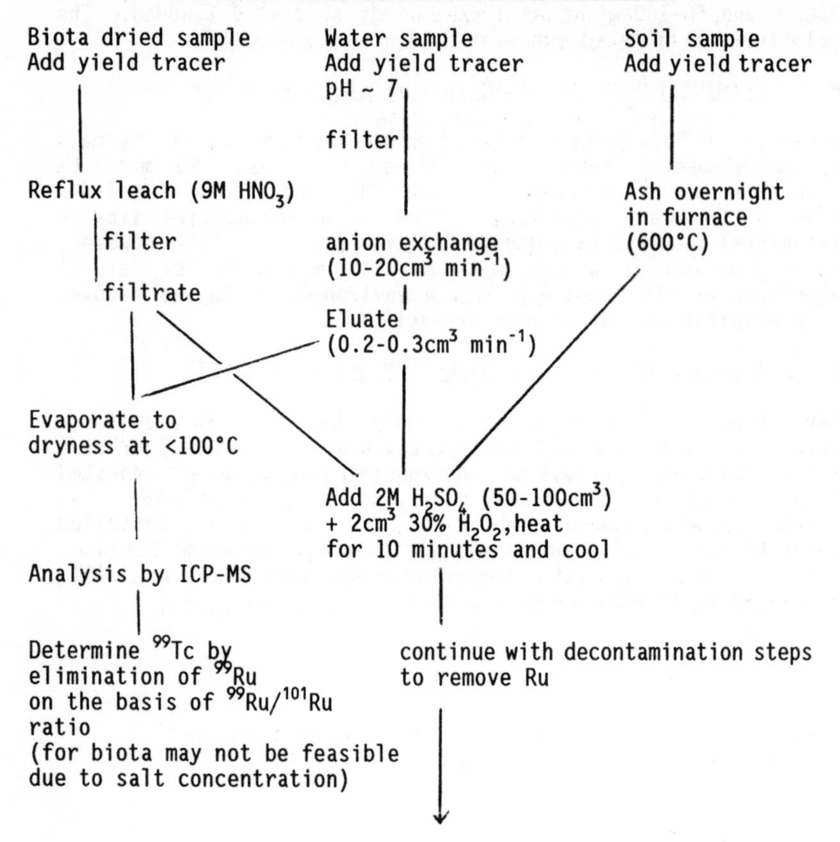

B. <u>Ruthenium decontamination steps</u>

Add 2M NaOH solution to ppt. $Fe(OH)_3$ at pH 12
Heat for 10 min, cool and filter.
Acidify to 2M H_2SO_4-NaOCl, boil for 30 min and cool

Add 1-2cm^3 5% NaOCl, extract with 2x50cm^3 CCl_4

Discard CCl_4

Acidify to 2M H_2SO_4-1M HCl (1-2cm^3 30% H_2O_2).
Extract with 50cm^3 cyclohexanone.
Wash organic layer with 50cm^3 4M H_2SO_4 (0.1cm^3 30% H_2O_2)

Discard
Aqueous and wash
phases

Add 50cm^3 cyclohexane and back -extract twice with 40cm^3 0.5M NaOH

Discard organic layer

Mix 1cm^3(10mg) $AgNO_3$ and 2cm^3 5% NaOCl. Heat for 15 min,
cool and filter

Filtrate

Make 2 M w.r.t. H_2SO_4 (1cm^3 30% H_2O_2), extract with 40cm^3
5% TIOA-xylene and wash the organic phase with 40cm^3
2M H_2SO_4 (few drops of 30% H_2O_2)

Discard aqueous
and wash phases

Back extract organic phase with 2x20cm^3 IM NaOH
Discard organic phase to C

↓

C Final preparation of technetium

Adjust pH to 7, then pass aqueous phase through
5cm^3 slurry of Dowex 1-x8 resin (50-100 mesh) (10-20cm^{3}min^{-1})

Discard filtrate

Wash the column with 100cm^3 cold and 100cm^3 hot distilled water

Discard washings

Elute Tc with 40-45cm^3 12M HNO_3 at a flow rate 0.2-0.3cm^3 mn^{-1}

Determine tracer yield by γ-counting

Evaporate the eluate to dryness

Add 5 or 10cm^3 of 2% HNO_3 and analyse ^{99}Tc by ICP-MS

3. RESULTS

% Chemical recoveries

The following chemical yields were obtained for various procedures on the different types of sample:

a) Water samples

1. Sample → anion exchange → Ru elimination → anion - exch → ICP-MS — 65±15% (adopted)

2. Sample → Ru elimination → anion - exch → ICP-MS — 75±20%

b) biota (algae) samples

1. Sample, wet (HCl), dry → ash (600°C) with fresh air → 3M H_2SO_4 → TIOA - xylene extraction →anion exchange → ICP-MS — 15%

2. Ash - fuse (Na_2O_2) → HCl → ppt(Ni, Fe) → anion exchange → cation exchange → cyclohexanone extraction → washing with chloroform → ICP-MS — 10%

3. Reflux/leach (9M HNO_3) → removal of Ru → anion exchange → ICP-MS — 70±11% (adopted)

c) Soils and sediments samples

1.Leach → Ru elimination → anion exchange → ICP-MS	20%
2.Ashing → Ru elimination → anion exchange → ICP-MS	65±15% (adopted)

Intercomparison Exercise (Studies)

The following reference materials were analysed in order to provide a check on the validity of the method:

1 IAEA/marine algae AG-B-1, obtained from IAEA, International Laboratory of Marine Radioactivity, Monaco.

2. Seaweed (*Fucus vesiculosus*);
Supplied by, R & D Department, BNFL plc, Sellafield, Cumbria, U.K.[13]

The results obtained in the present work are compared with certified values in Tables 1 and 2.

Limit of detection (L.O.D.)

For multiple runs on a sample, the mean and standard deviation are calculated for each isotope:

$$\text{Mean} = \frac{\Sigma(\text{sample counts})}{\text{runs}}$$

$$\text{Std Dev} = \sqrt{\frac{\Sigma(\text{sample counts - mean})^2}{\text{runs - 1}}}$$

The estimate of error for a given standard is given by:

$$\text{Error} = \frac{\text{Std Dev}}{\sqrt{\text{Runs}}}$$

$$\text{L.O.D.} = 3 * \sqrt{\text{error counts}}$$
$$= \underline{0.004} \text{ ppb}$$

Table 1 Intercomparison of Radionuclide Measurements in IAEA Marine Alga Sample AG-B-1.

Lab. Code No.	Tc-99, mBq/g
2	11.4 ± 0.5(2)
5	11.1 ± 1.5(5)
6	14.7 ± 0.2(2)
27	13.3 ± 1.2
28	11.5 ± 1.5
Overall median of 2,5&6 results, Md	11.5
Confidence interval (α = 0.05)	11.1-14.7
Our lab result	12.5 ± 1.4(5)

Table 2 Published results of BNFL intercomparison seaweeds.

Lab	gms taken	method	^{99}Tc(Bq/g,) Mean
1	2g	L.S. Spec:	4.54
2	2g	L.S. Spec:	4.42
3,A	2g	ß-counting	3.38
B	1g	Gas Flow C.	4.60
4	1g	ß-Counting	4.99
5	2-3g	ICP-MS	ca 4.0
6	2-3g	ICP-MS	<4.40
7	<1g	ICP-MS	4.70

The overall mean value obtained from laboratory means expressed as Bq/g is (1-6 samples):
4.64 ± 0.25 S.D. of a lab value
± 0.12 S.D. of the mean

(7) our laboratory result.

4 DISCUSSION

There are many procedures for the determination of ^{99}Tc. After a thorough search of the literature there was no published procedure suitable for the determination of ^{99}Tc levels in different types of environmental samples which paid adequate attention to the potential interference by ^{99}Ru using ICP-MS. Current literature for ^{99}Tc by ICP-MS is given in table 3.

Table 3 To date ICP-MS procedures for ^{99}Tc measurement.

Sample	Ru decontamination	Methods	L.O.D. (ppb)	Ref
N. Fuel Analysis	No	ICP	20	14
Aq. Samples	No	ICP-MS best from ICP-OES, IEX/LSC and GF AAS	0.04	15
Ag. Solutions	No	ICP-MS	0.01-0.02	16
Soil Samples	No	ICP-MS	0.0032	17
Soil Samples only	Yes	ICP-MS	0.01	18
Springfield, effluent	No	ICPMS, LSC	0.006	19

Because of the importance of and current interest in the collective long-term dose, effort has been made to optimise a procedure for the determination of ^{99}Tc which overcomes the high purity requirements of ICP-MS, and the low specific radioactivity found in environmental samples.

For the complete elimination of Ru isotope the established method[20] was applied with modification. The different valence states of Tc and Ru were controlled with H_2O_2 and NaOCl in different steps to achieve the most effective decontamination. The first step was the heating/boiling of solution at 100°C in 2M H_2SO_4 - NaOCl when RuO_4 was volatilized. In the second step RuO_4 was extracted by CCl_4 (+NaOCl) at pH=4 leaving TcO_4 in solution. This was followed by cyclohexanone and TIOA-xylene extraction. Precipitation steps were carried out to remove $Fe(OH)_3$ and AgCl.

REFERENCES

1. D.I. Mendeleev, Zh.Pf.Kho., 1869, 1, 60.
2. C. Perrier and E. Segre, J. Chemical Physics., 1937, 5, 712.
3. C. Perrier and E. Segre, Nature., 1947, 159, 24.
4. K.V. Kotegov, O.N. Pavlov and V.P. Shvedov., Advances in Inorg. Chem. and Radiochem., 1968, 11, 1
5. A. K. Lavrukhina and A.A. Pozdnyakov, Analytical Chemistry of the Elements, Ann Arbor-Humphrey Science, London, 1970.
6. J. Rioseco, ^{99}Tc, University of Lund, Sweden, 1987.
7. P Robb, PhD Thesis, University of Loughborough, 1983.
8. S.E. Long and S.T. Sparkes, AERE-R 12742, 1987.
9. E. Holm and J Rioseco, Nucl. Instr. and Methods in Physics Research, 1984, 223, 204.
10. J.E. Till, In: "Technetium in the environment", by Desmet & Mytcenaere, EUR 10102, Elsevier Applied Science Publishers, London and New York, 1986.
11. B.T. Kenna and P.K. Kuroda, J Inorg. Nucl. Chem., 1964, 26, 493.
12. Report No 27, IAEA/RL/129, Monaco, 1985.
13. T.H. Bates, Environment International., 1988, 14, 283.
14. K.D. Karnowski and B. Rolf, Laborpraxis., 1981, 5, 574.
15. R.M. Brown and C.J. Pickford, Analyst., 1984, 109, 673.
16. S.E. Long, R.M. Brown and C.J. Pickford, ICP Winter Conference, Lyon, France, 1987.
17. C. Kim, M. Otsuji, Y. Takaku, H. Kawamura, K. Shiraishi, Y. Igarashi, S. Igarashi and N. Ikeda, Radioisotopes., 1989, 38, 151.
18. S. Nicholson, T.W. Sanders, T. Cole and L.M.Baine, 6th International Symposium on Environmental Radiochemical Analysis, Manchester, 1990.
19. D.P. Bullivant, M. John and P.R. Makinson, 2nd International Conference on Plasma Source Mass Spectrometry, University of Durham, 1990.
20. Q. Chen, A. Aarkrog, H. Dick and K. Mandrup, J. Radioanal. Nucl. Chem., 1989, 131, 171.

The Determination of ^{99}Tc by ICP-MS in Samples Collected near Nuclear Installations

Ihsanullah and B.W. East

SCOTTISH UNIVERSITIES RESEARCH AND REACTOR CENTRE, EAST KILBRIDE, GLASGOW G75 0QU, UK

1. INTRODUCTION

^{99}Tc occurs naturally in the earth's crust primarily from spontaneous fission of ^{238}U and slow neutron-induced fission of ^{235}U[1,2]. ^{99m}Tc is prepared by the neutron irradiation of ^{98}Mo, and is widely used in nuclear medicine. ^{99}Tc is also produced from nuclear detonation tests, nuclear fuel processing plants and nuclear power stations. By comparison to other sources, the contribution to ^{99}Tc from natural sources and use of ^{99m}Tc in medicine is considered to be negligible[3,4].

Assuming that ^{99}Tc and ^{137}Cs are produced with the same representative fission yield the global activity of ^{99}Tc released into the stratosphere up until 1980 has been estimated to be 140 TBq. If fallout is also taken into account then a further total of 100 TBq of ^{99}Tc was released into the environment during 1945-1980[3]. The total core inventory of (^{99}Mo, the precursor of ^{99}Tc) in the Chernobyl Reactor accident (1986) was calculated by the Soviet experts to be 4.8 x 10^6 TBq while the percentage released to the environment was estimated to be 2.3 Percent of core inventory (1.1 x 10^5 TBq)[5].

Luykx[6] estimated the production of ^{99}Tc by nuclear power stations at the end of 1983 throughout the world was 15000 TBq (2400 kg) ^{99}Tc. This author estimated the global release to the environment from the nuclear fuel cycle to be of order of 1000TBq.

Releases from the nuclear fuel cycle include reactor operation, nuclear fuel reprocessing, UF_6 conversion, uranium enrichment, U fuel fabrication, high-level waste solidification, low and high-level waste disposal; however the main contribution to the release of ^{99}Tc is the process of uranium enrichment[3,6].

Luykx[7] estimates that the major discharges of ^{99}Tc in liquid effluents arise from nuclear fuel reprocessing plants. The Capenhurst enrichment plant in the UK had the following releases of ^{99}Tc: 1978 (2-6 GBq); 1979 (12.2 GBq); 1980 (11.5 GBq); 1981 (6.3 GBq); 1982 (20.4 GBq) and 1983 (3.4 GBq). The reprocessing plant at Cap de la Hague, France, released 11.7 TBq into the sea in 1983[3]. Discharges from the reprocessing and other activities at Dounreay Nuclear Power Development Establishment enter the Pentland Firth and the levels of radioactivity released to the sea are considerably smaller than those from the Sellafield and Cap de la Hague operations. Discharges of liquid effluents from the Sellafield reprocessing plant enter Irish Sea via pipelines which extend to 2.4 km below the low-water mark[8]. Figure 1 shows the discharges with liquid effluents from the Sellafield plant at Seascale, UK[7,9]. The values show an important decrease since 1981, the reason being that the previous years reflect delayed discharges from storage tanks[7] which were higher. Because of the ^{99}Tc activity found in the Irish Sea seaweeds and water samples have been investigated by various workers and ^{99}Tc concentrations found are given in Table 1.

In Figure 2 the results available for Fucus vesiculosus are given as Bq kg^{-1} (dry weight) of ^{99}Tc. The ^{99}Tc/^{137}Cs activity ratio is also given after the ^{99}Tc value separated by an oblique[18]. The present values (1982) are 15-60 times lower than the values obtained in 1978 ie 6-21 kBq kg^{-1} wet weight (30-100 kBq dry weight)[13]. This corresponds well with the decreased release of ^{99}Tc from 180 TBq in 1978 to 3.5 TBq in 1982.

^{99}Tc activity concentrations of 22-79 Bqkg^{-1} wet weight (~110-395 Bq kg^{-1} dry weight) in Fucus serratus at Cap de la Hague have been reported[14].

In the present work various environmental samples from the Irish Sea and from Chernobyl have been analysed for ^{99}Tc by newly developed procedures using ICP-MS[20].

2. EXPERIMENTAL

SAMPLING

The environmental samples ie, Fucus vesiculosus, Ascophyllum nodosum, water and Pyrophyra (sloke) samples were collected from different coastal sites (shown in Figure 3) from the Sellafield during 1989-1990. The Ravenglass silt was collected from the top 5cm of an area approximately 40cm x 40cm, 10 to 15m from the highwater mark.

Moss and lichen samples were collected some 1km east of the Chernobyl Nuclear Power Station in an area known as the "Forest of Miracles".

PROCEDURE

Due to the expected low specific radioactivity, the Tc from the environmental water sample was concentrated on anion-exchange resin from a bulk water sample. In order to obtain Tc in solution, Fucus vesiculosus, Ascophyllum nodosum, Pyrophyra (sloke), Moss and lichen samples were dried in oven (110°C) and were subjected to leaching. The silt sample was ashed. After initial sample treatment a procedure for Ru decontamination and the preparation of the samples in the required form for ICP-MS analysis was followed. The flow sheet of the procedure is given in Figure 4. The detailed procedure is to be published.

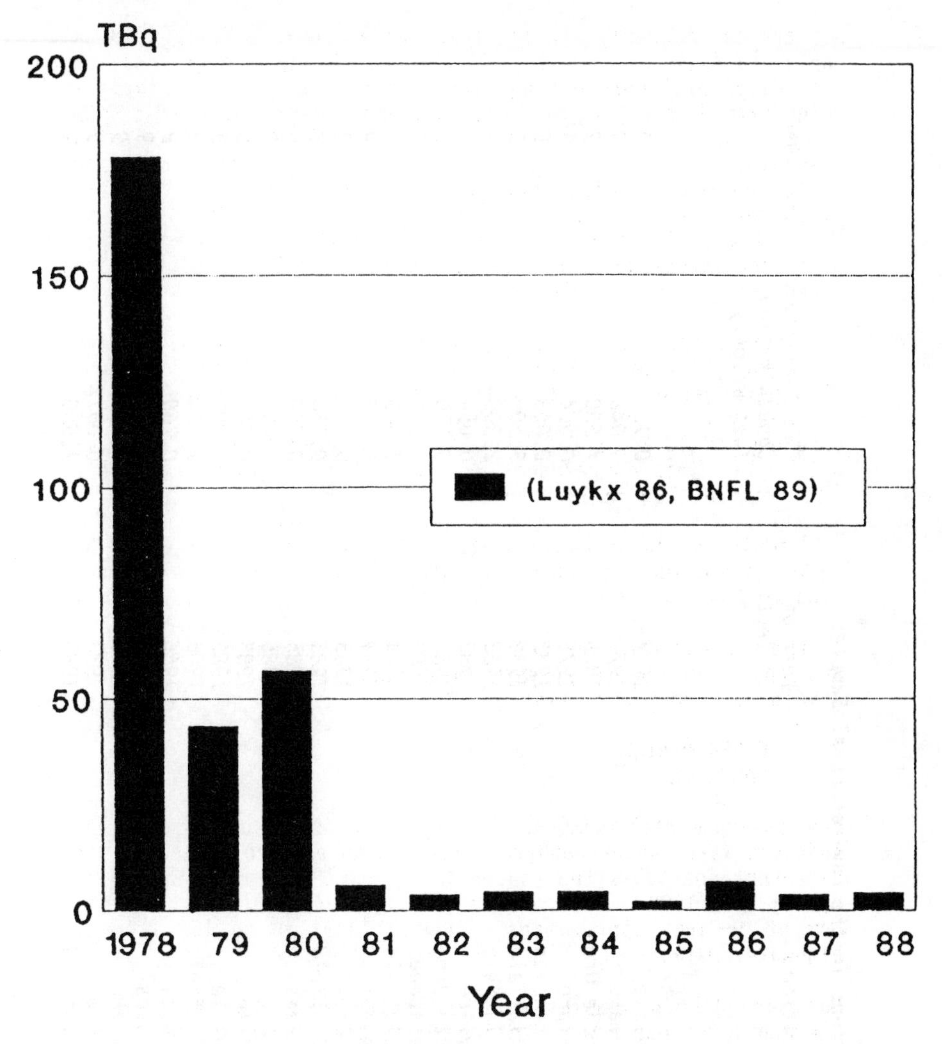

Fig:1 ^{99}Tc from Sellafield plant.

TABLE 1 ^{99}Tc Concentrations in the Environmental Samples

Samples	Location	Year of Collection	^{99}Tc Activity Concentration (wet weight) Bq g^{-1}	Ref
Fucus vesiculosus	Hoylake	1977	0.120	9
Fucus spiralis	Hoylake	1977	0.048	9
Fucus vesiculosus	Little Orme	1978	4.500	9
Sand	Hoylake	1978	0.004	9
Silt	Whitehaven	1982	0.008	9
Fucus vesiculosus	Sellafield	1982	3.500	9
Fucus vesiculosus	Sellafield	1983	2.300	9
Fucus spiralis	Hoylake	1983	0.054	9
Fucus spiralis	Litte Orme	1983	0.310	9
Prophyra	Braystones	1984	0.012	9
Fucus vesiculosus	Sellafield	1984	1.800	9
Fucus vesiculosus	Heysham	1984	0.360	9
Porphyra	St Bees	1985	0.0015	9
Fucus vesiculosus	Sellafield	1985	0.750	9
Fucus vesiculosus	St Bees	1985	0.710	9
Ascophyllum nodosum	St Bees	1985	1.700	9
Fucus vesiculosus	Little Orme	1985	0.120	9
Silt	Hoylake	1985	0.00255	9
Fucus vesiculosus	Sellafield	1986	1.200	9
Fucus vesiculosus	St Bees	1986	0.760	9
Ascophyllum nodosum	St Bees	1986	1.000	9
Silt	Hoylake	1986	0.005-4	9
Porphyra	St Bees	1988	0.0017	9
Fucus vesiculosus	Sellafield	1988	1.900	9

TABLE 1 Continued

Samples	Location	Year of Collection	^{99}Tc Activity Concentration (wet weight)	Ref
Fucus vesiculosus	St Bees	1988	0.720	9
Sea water	Irish Sea	1969	0.0045 ± 0.009	10
Sea water	Irish Sea	1972	0.108 ± 0.007	11
Sea water	North Sea	1980	0.007 ± 0.001	19
Sea water	Mediterranean Coastal Sea water	1985-86	~0.00007 ± 0.00003	19
Rain	Texas	1971	$2.4\ 10^{-4}$	17
Seawater	Coastal Irish Sea	1982	0.015 - 0.074	15
Fucus vesiculosus	Coastal Irish Sea	1982	0.66 ± 0.08*	15
Fucus vesiculosus	St Bees	1973	16.28	12
Porphyra	St Bees	1973	0.026	11
Fucus vesiculosus	Irish Sea	1978	4.55 - 21.28	13
Silt	Ravenglass	1978	0.009	13
Ascophyllum nodosum	Southern Brittany	1977	0.126	14
Fucus vesiculosus	Southern Brittany	1979	$<7.4 \times 10^{-4}$	14
Sea water	Irish Sea	1982	0.015 - 0.074	11
Fucus vesiculosus	Irish Sea	1985-86	0.0014 - 0.410*	16

*Biota and silt = Bq g^{-1} (dry weight)
water = Bq l^{-1}

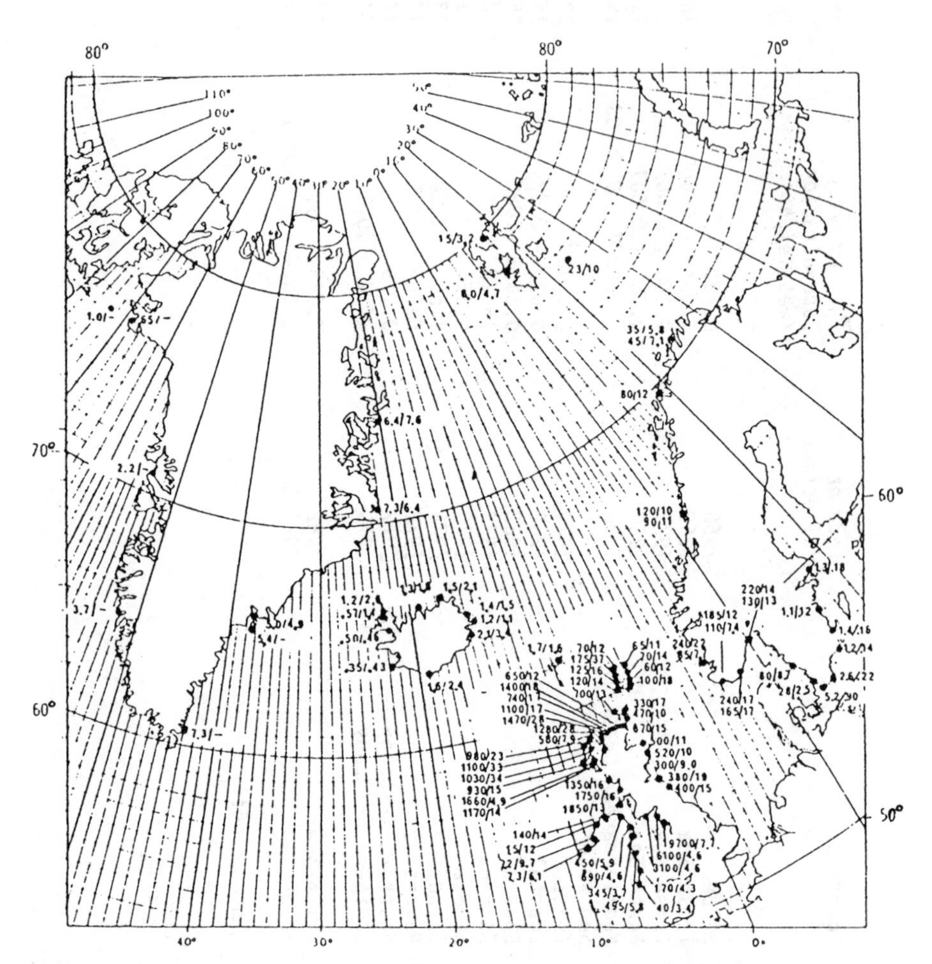

FIG. 2. Activity concentration of ^{99}Tc in *Fucus vesiculosus* (Bq kg^{-1} dry weight) from the North Atlantic and the Arctic Ocean. For the Faroes and Iceland results for *F. disticus* were also used. Results are from the Faroes and Iceland in 1981, from Norway in 1981 and 1982 (below), from Sweden in 1982, from England and Scotland in 1982, from Ireland in 1983 and from Greenland during 1979–82. The ^{99}Tc/^{137}Cs activity ratio is given after the solidus. (18)

Fig:3 UK nuclear establishments (•) and sampling sites (o)

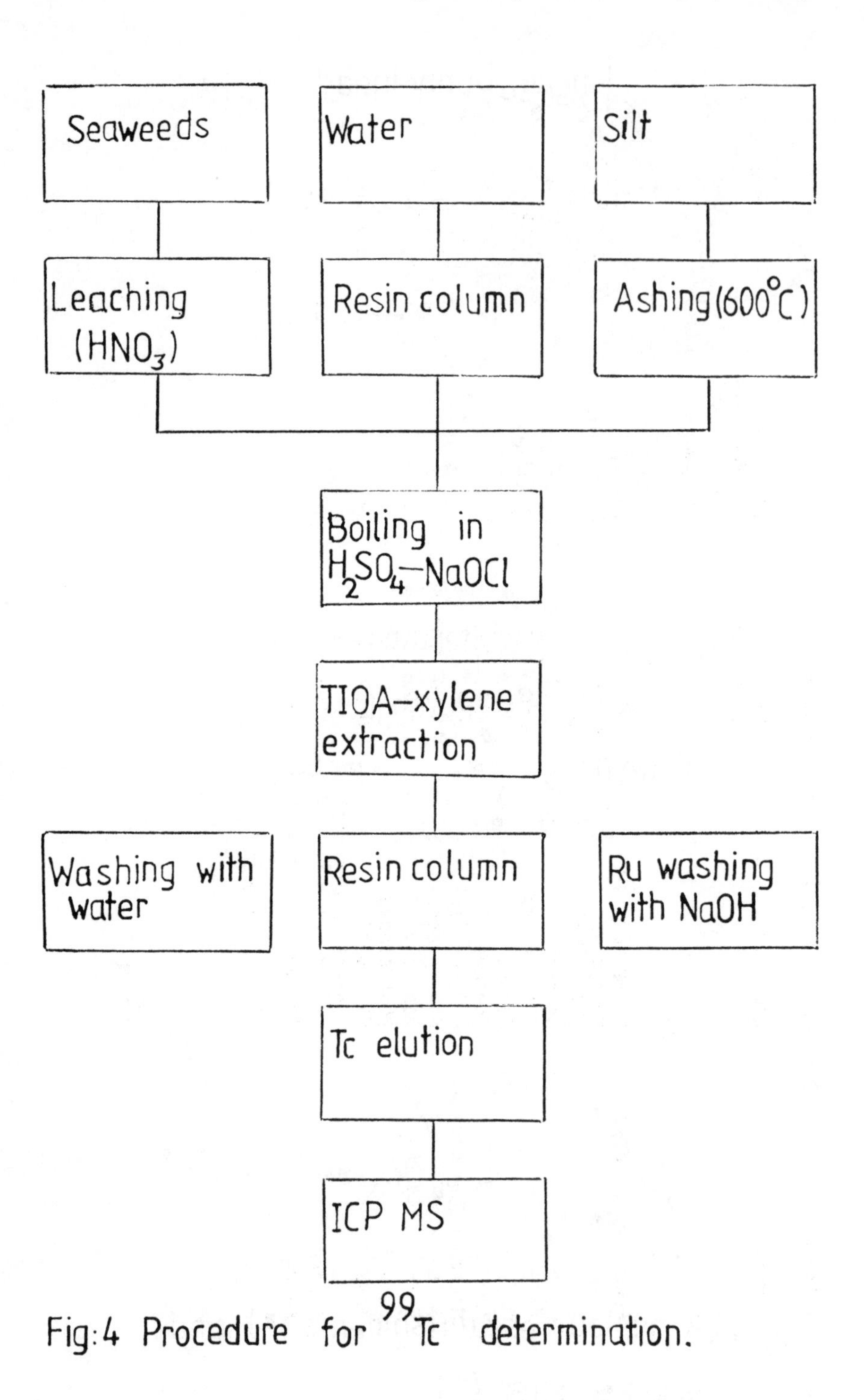

Fig: 4 Procedure for ^{99}Tc determination.

TABLE 2 ^{99}Tc **Concentration in Environmental Samples**

Sample	Location	Year of Collection	Quantity for Analysis (dry weight) (in gms)	^{99}Tc (Bq g^{-1}) ± δD
Fucus vesiculosus	Colwyn Bay	9/89	24	0.74 ± 0.05
" "	Ravenglass	10/89	25	10.18 ± 0.53
" "	Whitehaven	10/89	23	1.98 ± 0.06
" "	Sandyhill	10/89	24	1.78 ± 0.09
" "	Ettrick Bay	4/90	25	0.49 ± 0.04
" "	Tongue	4/90	24	0.14 ± 0.02
" "	Aberdeen	4/90	25	0.10 ± 0.05
Ascophhyllum nodosum	Ettrick Bay	4/90	20	1.06 ± 0.31
Silt	Ravenglass	11/86	30	0.025 ± 0.001
Porphyra (Sloke)	Dounreay (Dunnethead)	7/90	5	$<2.5 \times 10^{-3}$
Moss	Chernobyl (Forest of Miracle)	10/89	1	$<12.6 \times 10^{-3}$
Lichen	" "	10/89	0.2	$<63.0 \times 10^{-3}$
Water	Tongue	3/90	7.5L	<17 Bql^{-1}

RESULTS

All the samples were analysed in duplicate with the results given in Table 2.

4. DISCUSSION

As described, various types of sample were analysed in order to verify new ICP-MS procedures and to determine environmental ^{99}Tc levels. Fucus vesiculosus was selected for many sites as it is well known that this species of brown algae is an excellent bioindicator for various radionuclides including technetium[18]. The concentration factors (CF) for freshwater and marine brown algae are very high between 250 and 2500 while for green algae they are only 1-10 and for red algae similarly 1-20[21]. In some cases sampling of species other than Fucus vesiculosus was performed. It would of course have been desirable to use several species as bioindicators and obtain comparative results. For example, the ratio for Ascophyllum nodosum/Fucus vesiculosus (2.16) is in good agreement with the ratio in the literature i.e. 2.1 ± 0.4[18] and ^{99}Tc concentration in Poryphyra relative to Fucus vesiculosus[17] is $(1.8 \pm 0.28) \times 10^{-3}$. Other comparisons could therefore have been useful but were not possible in this study.

The lichen and moss samples from Chernobyl were from an undisturbed area and because of their high radiocaesium content were assumed to have been present at the time of the reactor accident. Their ^{99}Tc content is thus probably representative of that present in fallout in the area in the vicinity of the reactor.

The recovery of Tc from the Ravenglass silt samples as measured by ^{95m}Tc tracer appeared satisfactory when compared with the biota samples.

5. CONCLUSIONS

Our results represent some of the first recorded ^{99}Tc data for seaweeds and other environmental samples by ICP-MS analysis. Ascophyllum nodosum is found to concentrate technetium twice as efficiently as Fucus vesiculosus and Pyrophyra is poor bioindicator. The results for Fucus vesiculosus are in good agreement with MAFF reports.

ACKNOWLEDGEMENT

The authors are grateful to Prof M S Baxter, Dr P MacDonald and Mr Keith McKay for their help and guidance. One of the authors (Ihsanullah) is also grateful to the Ministry of Science and Technology, Pakistan and the Pakistan Atomic Energy Commission for a scholarship and leave of absence respectively, which made this work possible.

REFERENCES

1. D. B. Curtis, J. H. Cappis, R. E. Perrin and D. J. Rokop, Applied Geochemistry, 1987, 2, 133.

2. B. T. Kenna and P. K. Kuroda, J. Inorg. Nucl. Chem., 1964, 26, 493.

3. J. Rioseco. "^{99}Tc", University of Lund, Sweden, 1987.

4. K. C. Ehrhardt and M. Jr. Attrep, Environ. Sci. Technol., 1978, 12, 55.

5. INSAG-I Report, p.34.

6. J. E. Till, In: Technetium in the environment, by Desmet and Myttenaere, EUR 10102, Elsevier Applied Science Publishers, London and New York, 1986.

7. F. Luykx, In: Technetium in the environment, by Desmet and Myttenaere, EUR 10102, Elsevier Applied Science Publishers, London and New York, 1986.

8. W. C. Camplin and A. Aarkrog, Fisheries Research Data Report No. 20, Lowestoft, 1989.

9. J. Hunt, Aquat. Environ. Monit. Rep., MAFF Direct. Fish. Res., Lowestoft, 1979-1989.

10. N. W. Golchert and J. Sedlet, Anal. Chem., 1969, 41, 669.

11. E. Holm, J. Rioseco, S. Ballestra and A. Walton, J. Radioanal. Nucl. Chem. Articles, 1988, 123, 167.

12. J. W. R. Dutton and R. B. Ibbett, Symposium (2-3 April, 1973) London.

13. R. J. Pentreath, D. F. Jefferies and M. B. Lovett, In: Proceedings of 3rd NEA Seminar, Tokyo, Japan, 1979, 203-221.

14. L. Jeanmarie, M. Masson, F. Patti, P. Germain and L. Cappellini, Mar. Poll. Bull., 1981, 12, 29.

15. J. P. Riley and S. A. Siddiqui, Analytica. Chimica. Acta., 1982, 139, 167.

16. M. Garcia-Leon, J. of Radioanaly. Nucl. Chem. Chem. Articles, 1990, 138, 171.

17. M. Attrep, J. A. Enochs and L. D. Broz, Envir. Sci. Technol., 1971, 5, 344.

18. E. Holm, J. Rioseco, A. Aarkrog, H. Dahlgaard, L. Ballstadius, B. Bjurman and R. Hedvall, In: Technetium in the Environment, by Desmet and Myttenaere, EUR 10102,

Elsevier Applied Science Publishers, London and New York, 1986.

19. S. Ballestra, G. Barci, E. Holm, J. Lopez and J. Gastand, J. Radioanal. Nucl. Chem., Articles, 1987, 115, 51.

20. Ihsanullah and B. W. East, (will be published later).

21. B. G. Blaylock, M. L. Frank, F. O. Hoffman and D. L. Dengelis, In: Technetium in the environment. by Desmet and Myttenaere, EUR 10102, Elsevier Applied Science Publishers, London and New York, 1986.

Some Observations on Mass Bias Effects Occurring in ICP-MS Systems

P.J. Turner

TURNER SCIENTIFIC, UNIT 7, ASHER COURT, LYNCASTLE WAY, APPLETON, WARRINGTON WA4 4ST, UK

1 INTRODUCTION

Large and variable mass bias effects have commonly been observed in many inductively coupled mass spectrometer systems in use at the present time.[1] The general form of this bias is a severe loss of transmission for light elements or a droop in the transfer characteristic at both ends of the normal operating mass range (Li-U) which may be varied by adjusting the operating parameters of the system. This is obviously an unsatisfactory situation in which to carry out analytical measurements, since small changes in sample or instrument state can introduce significant deviation from a previously measured calibration curve. These biases are essentially electrostatic in origin. Any transfer lens system will have focal properties which depend on the energy of the particles passing through the system. There will usually be one particular energy for which the system is well focussed providing efficient transport through a series of apertures, whilst beams of different energies will be focussed at different points in the system. This will result in a reduction in transfer efficiency for particles having energies different from the optimum. This effect is large in ICP MS since particles travel through the sample and skimmer interface apertures with an approximately constant velocity independent of mass.[2] This results in a energy which is approximately proportional to mass.

A further source of mass discrimination arises from so called "space charge" effects behind the skimmer cone. Here, there is a substantial column of positively charged particles travelling towards the ion optical elements of the mass spectrometer. Each positively charged particle is repelled by the other positively charged particles with the normal coulomb force $\frac{e^2}{r^2}$. Transverse acceleration is therefore proportional to $\frac{e^2}{m\,r^2}$ so that lighter particles will tend to be lost more rapidly from the beam than will heavy ones. If the transit time into the ion optics is large compared with the time required for the lighter ions to acquire substantial lateral velocity, there will be significant mass bias introduced into the system.

Both these effects may be minimised by the use of conventional beam transport principles.

2 LAYOUT OF ANALYSER SYSTEM

A three aperture sampling system has been used in the TS SOLA plasma-to-mass spectrometer interface region. This is shown in figure 1. The first two apertures are in the sampling cone and skimmer cone as described in various publications,[1] whilst the third aperture is in the accelerator cone which is some 10mm behind the skimmer cone. This arrangement is identical to that commonly used in duoplasmatron ion sources [3] and has been described by Hayhurst and Telford [4] in their plasma sampling system. The main function of this electrode is to provide a strongly accelerating and convergent electrostatic field to minimise the space charge spreading of the substantial ion beam emerging from the plasma sampling interface. A secondary function of the accelerator cone is to act as a differential pumping aperture between the second and third pumping stages of the vacuum system. The typical operating pressures are 2 torr for the expansion chamber, 5×10^{-4} torr for the skimmer cone-acceleration cone region and 2×10^{-5} torr for the transfer lens and quadrupole region. With this arrangement, the distance travelled by the ion

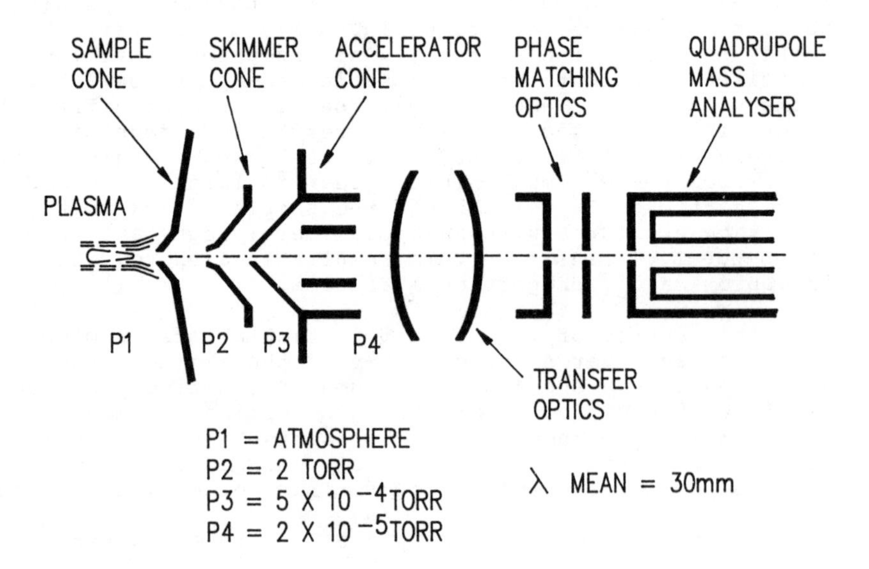

Figure 1 General layout of analyser system

beam in the relatively high pressure region of the second pumping stage is minimised, with a corresponding minimisation of the ion beam scattering losses.

An accelerating potential of 2000 volts is used on the acceleration cone, and behind this a simple x,y deflection system with a single variable einzel lens focussing potential is used to transfer the ion beam onto the entrance aperture of the analysing mass spectrometer. By using a transfer lens system operating at relatively high potential it is possible to focus ions of a wide energy spread into the quadrupole, and hence minimise the mass discrimination in this section of the instrument.

Figure 2 shows the manner in which the acceleration section of the ion optics function as modelled by SIMION.[5] It is possible to produce a fine cross-over in the centre of the cone for a wide spread of energies and angles of particles emerging from the sampling interface. This permits a good differential vacuum to be obtained between the second and third stages of the vacuum system whilst maintaining high ion beam transmission. A further advantage of the system is that it may be used as an energy filter simply by varying the total accelerating voltage on the acceleration element.

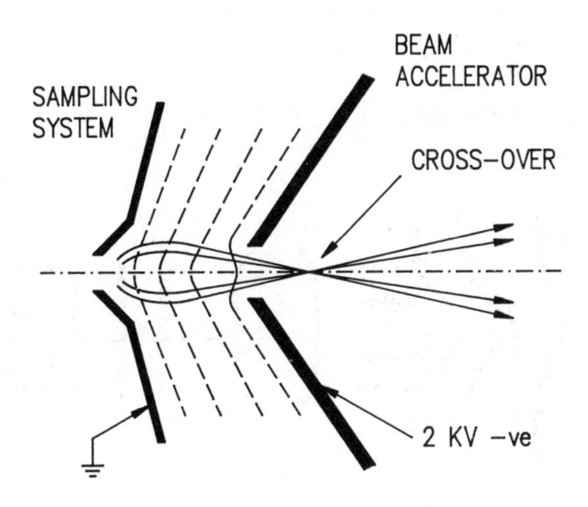

Figure 2 Ion beam trajectories in the region of the accelerator electrode

3 MEASUREMENT OF MASS BIAS CHARACTERISTICS OF ANALYTICAL SYSTEM

The transmission of the system was measured for elements of various atomic masses using solutions prepared from standards obtained from B D H Ltd.[6] Typical values are set out in table 1.

Table 1 Measured elemental sensitivites

Element	Mass	Ions/sec/ppm wt.
Be	9	$1.6x10^8$
Mn	55	$6.0x10^7$
Co	59	$6.8x10^7$
Sr	88	$3.6x10^7$
In	115	$3.4x10^7$
Pb	208	$1.2x10^7$

Sensitivites have been expressed as ions per second arriving at the collector per part million weight. In this case a Faraday collector was used and instrument tuning was carried out on the $^{115}In^+$ peak. A standard Meinhard type concentric pneumatic nebuliser was used for introduction of the sample into the system.

It can readily be seen that using this conventional mode for expressing sensitivities, lighter elements are substantially more sensitive than heavier ones.

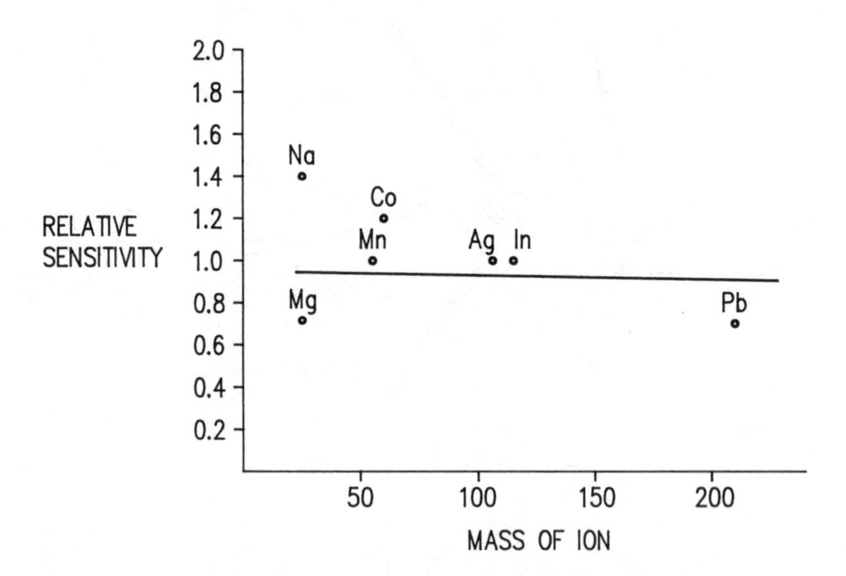

Figure 3 Relative sensitivity expressed in terms of atomic concentrations

It is more useful to express sensitivities in terms of atomic concentrations rather than weight concentrations to show mass biases. This has been done in figure 3.

The sensitivities in terms of atomic concentrations have been normalised to that of In at mass 115. It can be seen that instrument transmission is almost independent of mass, and that this style of ion optics has a truly "flat" characteristic. For semi-quantitative analyses a single transmission constant may be used for atomic concentrations, and this may be converted to a weight concentration simply by multiplying by relative atomic weights.

The elements shown in figure 3 are "well behaved" elements which are assumed to undergo 100% ionisation in the plasma flame. Other elements may be assumed to have the same mass transmission efficiency with relative sensitivities which are determined by their degree of ionisation in the plasma. The relative values for a number of high ionisation potential elements are shown in table 2.

Table 2 Normalised relative sensitivities for high ionisation potential elements

Element	Ionistion Potential	Normalised Relative Sensitivity
Zn	9.39	.59
As	9.8	.21
Se	9.75	.20
I	10.4	.15

Freedom from mass bias has several beneficial analytical consequences. Matrix effects are reduced, and whilst absolute sensitivities may be reduced in the presence of, for example, high concentrations of NaCl, relative sensitivities are not affected.

The effect of increasing concentrations of NaCl on the sensitivities of Mg, In and Pb is shown in figure 4. With the addition of .25 wt% NaCl, sensitivities have dropped by about 50%. Relative sensitivities remain constant, so that compensation by the use of an appropriate internal standard can be carried out.

Medium term stability is typically $\pm$ 2% over 30 mins., and sensitivity drift may be compensated across the full mass range using only one internal standard. A partial set of results using Sc as an internal standard is given in table 3. The data for table 3 was obtained from a mixed standard containing 25 elements at a concentration of 1ppm wt. After calibration, repeat measurements were carried out at intervals of 10 mins., and Sc was used as an internal standard.

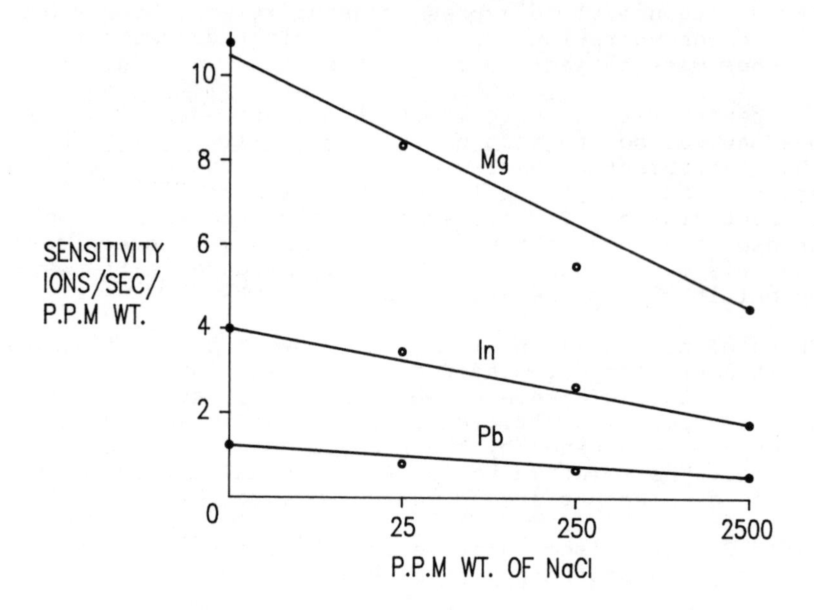

Figure 4 Effect of addition of NaCl on elemental sensitivities

A third significant benefit arising from a response curve which is flat in terms of atomic concentrations is the absence of the need to apply bias factors to isotope ratio measurements.

Figure 5 shows a schematic flat response and a schematic curved response. If the true relative intensity of ion beams measured at masses m_1 and m_2 is to be calculated, it is normally necessary to apply the function $f(m_1,m_2)$ to the observed ratio. This can in some cases be a relatively rapid function of m_1 and m_2 and depend quite sensitively on the tuning of the instrument. Moreover, small changes in lens potentials, sampling system etc., can also have an

Table 3 Variation of apparent concentration as function of time

Element	Meas.1	Meas.2	Meas.3	Meas.4	$\bar{x}$	σn
Li	.991	1.04	1.05	1.08	1.04	.03
Mg	1.02	1.09	1.02	1.02	1.04	.03
Al	1.06	1.12	1.08	1.05	1.05	.03
Mn	1.01	1.05	1.03	1.04	1.03	.01
Cu	.997	1.03	1.03	1.07	1.03	.03
Mo	1.02	.988	.996	.977	1.00	.02
Sn	.998	1.01	.976	1.04	1.01	.02
Er	1.00	1.06	.976	1.05	1.02	.03
Pb	.990	1.04	1.01	1.05	1.02	.03
U	.984	1.06	1.00	1.o6	1.03	.03

Figure 5 Schematic response curves for ICP MS systems

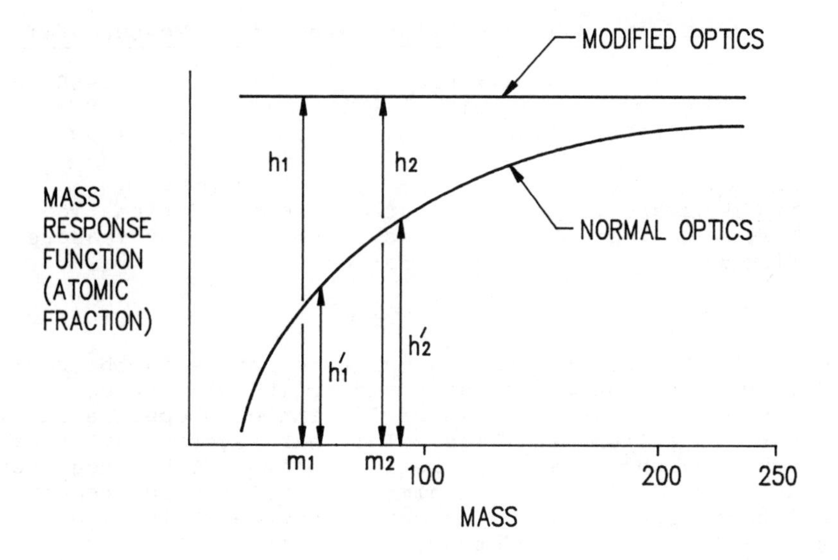

$$\frac{\text{true } c(m_1)}{\text{true } C(m_2)} = \frac{\text{meas } C(m_1)}{\text{meas } C(m_2)} \times f(m_1, m_2)$$

$$f(m_1,m_2) = \frac{h'_2}{h'_1} \qquad \text{for normal optics}$$

$$= \frac{h_2}{h_1} = 1 \qquad \text{for modified optics}$$

effect on the curve during the course of a measurement and this can have a disturbing effect on isotope ratio results.

For a flat response curve, any instabilities tend merely to change absolute sensitivities whilst relative intensities are unaffected. Isotope ratios are therefore stable and free from bias.

Values reported in table 4 were obtained using the Faraday collector. No biases have been applied to the results, and over the full mass range from Li to U, ratios are within 1% of the "true" value. This is particularly interesting for the Li isotope ratio measurement, where the relative mass difference between the isotopes is 15%. This causes severe problems in thermal ionisation isotope

Table 4 Isotope ratios obtained from ICP MS with low bias optics

Element	Ratio	Measured Value	Precision	Measured/Time
Li	6/7	8.031×10^{-2}	.35%	.988
Zn	66/64	5.675×10^{-1}	.31%	.998
U	235/238	1.007	.23%	1.007

ratio mass spectrometers, where isotope fractionation effects can lead to deviations of up to 5% for measured ratios with respect to true ratios.

4 SUMMARY

The use of conventional ion gun type optics in the region following the sampling interface of ICP MS systems produces an arrangement which is simple to operate and is almost free from mass bias. This has several analytical advantages when measuring real samples. It has been found that matrix effects are minimised, calibrations remain constant with time, and measured isotope ratios are typically within 1% of true values without the necessity for correction over the full working range of the instrument.

REFERENCES

1. G. Horlick et al in Inductively Coupled Plasmas in Analytical Atomic Spectrometry. Editors: A. Montaser, D.W. Golightly. V.C.H. Publishers ISBN 0-89573-334-x
2. D.J. Douglas, J.B. French, J. Anal. Atom. Spectrom 3, Sept 1988.
3. C.J. Lejeune, Nuclear Instruments and Methods, 1974 116, 417-428
4. A.N. Hayhurst, N.R. Telford, Combustion and Flame, 1977, 28, 67-80.
5. SIMION. D.A.Dahl, J.E.Delmore. Idaho National Engineering Laboratory, E.G.&G. Idaho Inc., P,O. Box 1625, Idaho Falls, ID 83415.
5. B.D.H. Ltd., Broome Road, Poole, Dorset, BH12 4NN.

Diagnostic Investigations on Ion Formation in ICP-MS

N. Jakubowski, I. Feldmann, and D. Stuewer

INSTITUT FÜR SPEKTROCHEMIE UND ANGEWANDTE SPEKTROSKOPIE, POSTFACH 10 13 52, D-W4600 DORTMUND 1, GERMANY

1 INTRODUCTION

In analytical research, the primary aim is to improve the methodological figures of merit, such as detection limits, and reliability or economic aspects - in order to supply analytical methods for the interdisciplinary demand of today and tomorrow[1]. Applying this general principle to ICP-MS, it is - despite its wide-spread acceptance - still a rather young technique offering many chances for further improvement of figures of merit. This refers not only to an enhancement in detection power, but to the minimization of spectral interferences and likewise matrix effects. Problems of this type, the origin of which is localized in the plasma or the interface region in most cases, have neither been sufficiently investigated nor fully comprehended, so that further basic research into atomization, ionization and molecule formation in the plasma and interface region is necessary. Investigations of such processes in the plasma have often been carried out with various spectroscopic methods. In the investigations presented here, the ICP-MS equipment itself is actually used as a direct plasma probe for diagnostic investigations on ion formation in the ICP.

An effective and easily available method to improve the efficiency of an ICP-MS instrument is the optimization of the operational parameters, as described in detail by various authors. In most cases however, these papers do not include comprehensive explanations of the processes in the plasma, which are influenced by such optimization efforts. It is the aim of these investigations to contribute to a better understanding of the influences which operational parameters may have on the formation of analyte and molecular ions in the plasma. For this purpose, spatial distributions of individual ion species are investigated as well as their energy distributions.

In ICP-MS, measurement of the spatial ion distribution is possible with adequate resolution. This is confirmed by a model of Douglas[2]. Although ions are collected from a volume the diameter of which is 8 times greater than the diameter of the sampler, a considerably higher spatial resolution is realized, because the skimmer and not the sampler limits the gas stream passed to the ion lens system, and this provides a resolution of better than 0.5 mm.

For these investigations, a highly flexible 40 MHz ICP-MS instrument developed by us has been used[3]. This instrument has been built on the basis of experiences gathered by the preceding development of a glow discharge instrument for the direct analysis of conducting solid samples mainly making use of the same hardware[4]. It includes a 40 MHz ICP generator with a coil configuration and a corresponding matching-unit also developed by the authors. A step motor is used for XYZ-positioning of the torch relative to the sampling cone, which enables measurement of the spatial ion intensity distributions. The diameter of the sampling orifice is 1 mm and that of the skimmer is 0.7 mm. The differential pumping system is made up of turbomolecular pumps. The detector system combines counting and analog mode. The analog mode has always been used here in order to register the intensity by a x-y-recorder. The equipment includes a mass flow controller. Furthermore, the whole

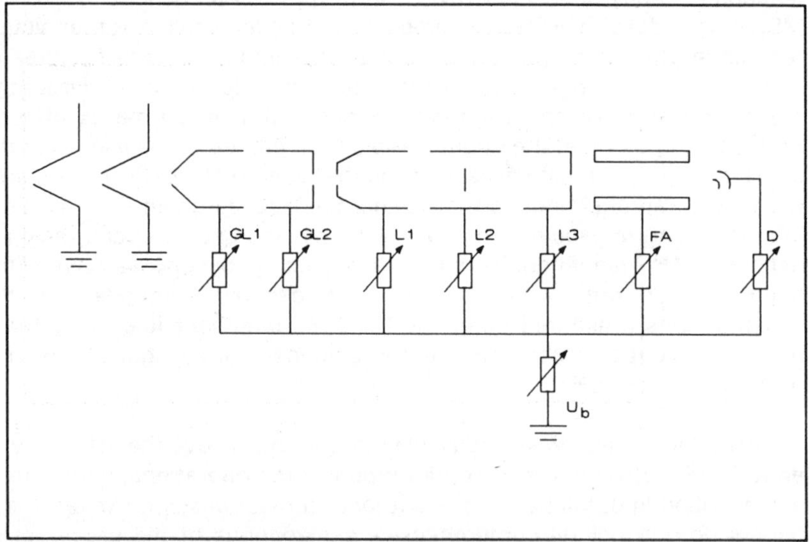

Fig. 1 Ion optical arrangement with bias potential technique: GL1, GL2 - grid lenses; L1-L3 - energy analyzer; FA - field axis; D - deflector; U_b - bias potential.

ion optical arrangement is connected to a voltage supply defining a common bias potential U_b as shown in Fig. 1. Variation of this potential yields a representation of the ion kinetic energy distribution[5], the so-called ion energy characteristic (IEC). This is an effective means to get an insight into ion formation in the plasma as has also been emphasized recently by Vickers et al.[6].

For these investigations, a GMK-nebulizer was used, the properties of which have been characterized in a previous paper[7]. Approximately 40 % of the aerosol droplets generated by this type of nebulizer have a diameter of 6 μm or below. The efficiency of the nebulizer is about 1 %. If not mentioned otherwise, Cu has been used as analyte element.

2 INFLUENCES OF OPERATIONAL PARAMETERS

Influences of the aerosol gas flow rate (AF) will be considered first. The AF effects several operational parameters during pneumatic nebulization, namely the nebulization efficiency, the droplet size distribution, the transportation properties, the speed of the aerosol particles in the plasma, and finally also the temperature of the plasma in the aerosol channel which is considerably lower than the temperature of the enveloping outer plasma[8].

In order to investigate the influence exerted by the AF on the ion formation the ion intensity was recorded as a function of the sampling

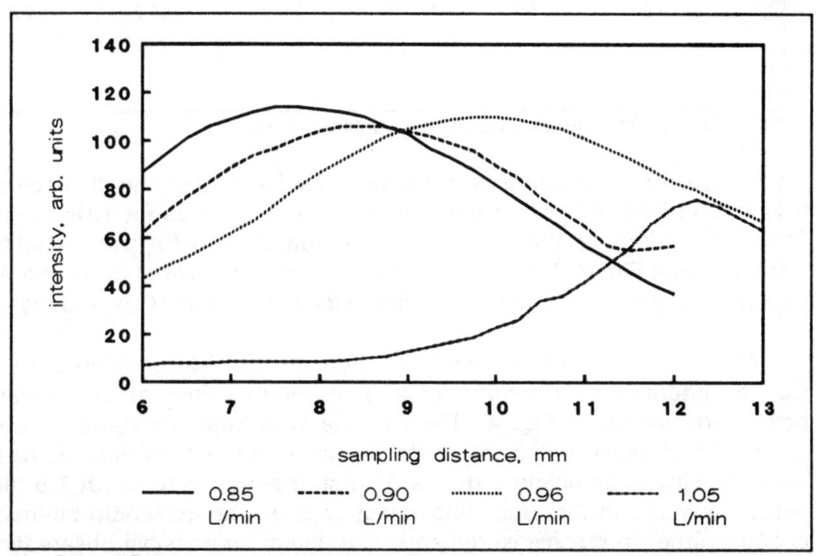

Fig. 2 Axial Cu^+ ion intensity distribution for different aerosol gas flows.

distance, i.e. the distance of the sampling cone from the load coil. Fig. 2 shows the axial ion intensity distribution obtained for a 1 ppm Cu-solution with a power of 1 kW as a function of the sampling distance for four different values of the AF. Intensity maxima occur between 7 mm and 12 mm. The maximum value of the intensity remains constant over wide variation range. With increasing AF, the location of atomization and ionization is shifted to a greater sampling distance which must be ascribed to the higher speed of the aerosol droplets. An increase of the AF by 0.05 L/min leads to an axial shift of 1 mm for the intensity maximum. This demonstrates that the sampling distance is highly sensitive to optimization, but it must be suspected that in the practice almost nobody takes heed of it.

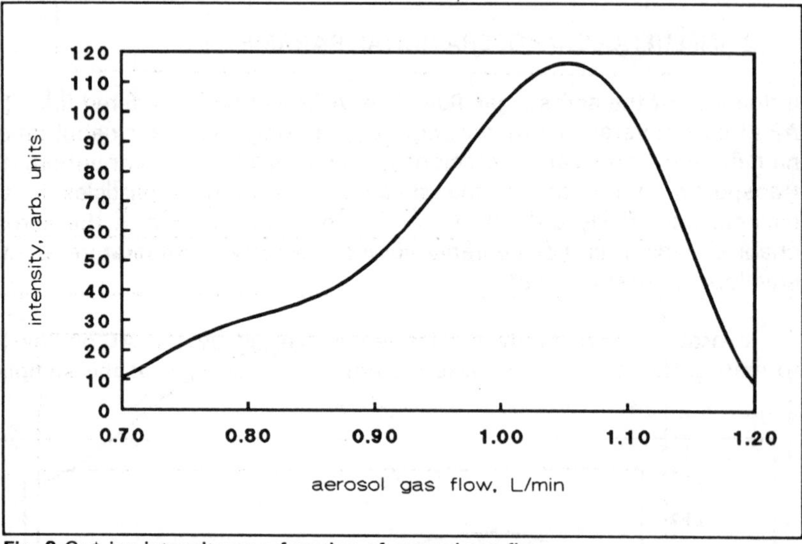

Fig. 3 Cu^+ ion intensity as a function of aerosol gas flow

Fig. 3 shows the intensity of Cu ions at a fixed sampling distance as a function of the AF with a pronounced maximum at about 1.05 L/min. This corresponds to the trend which can be derived for this sampling distance from Fig. 2. It should be emphasized that a variation of the AF by a few tenths L/min can result in intensity changes of 10 % and more.

Next, the influence of the electrical power will be considered. Axial Cu^+ ion intensity distributions for three different values of the forward power are shown in Fig. 4. The AF rate was kept constant at 0.96 L/min. An increase in power shifts the maximum of the ion intensity towards a lower sampling distance. This shift amounts to about 1.5 mm per 0.1 kW. As in the preceding investigations, the maximum intensity changes little. In our measurements, the minimum possible observation height is 6 mm above the load coil. For high power values, the intensity maximum is shifted nearer to the coil region than can be observed, as it is the case here for a power above 1.5 kW.

As an example of the analog representation of the ion intensity as a

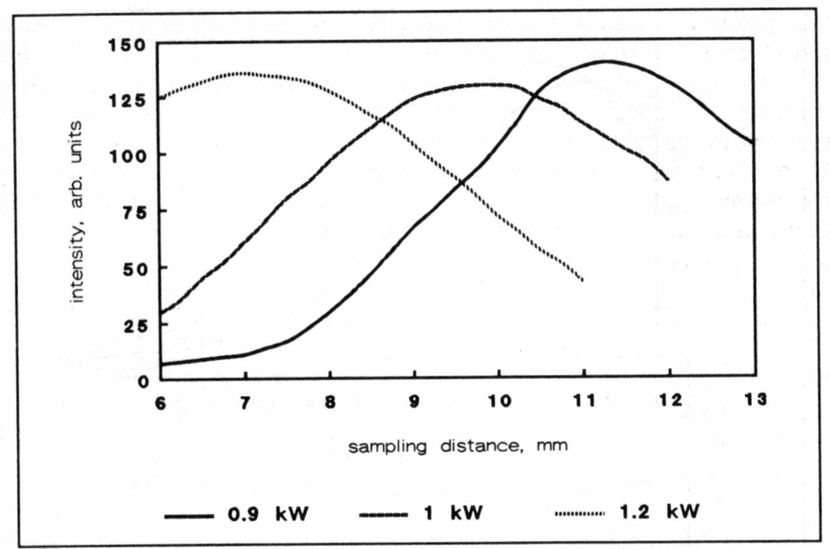

Fig. 4 Axial Cu^+ ion intensity distribution for different forward powers.

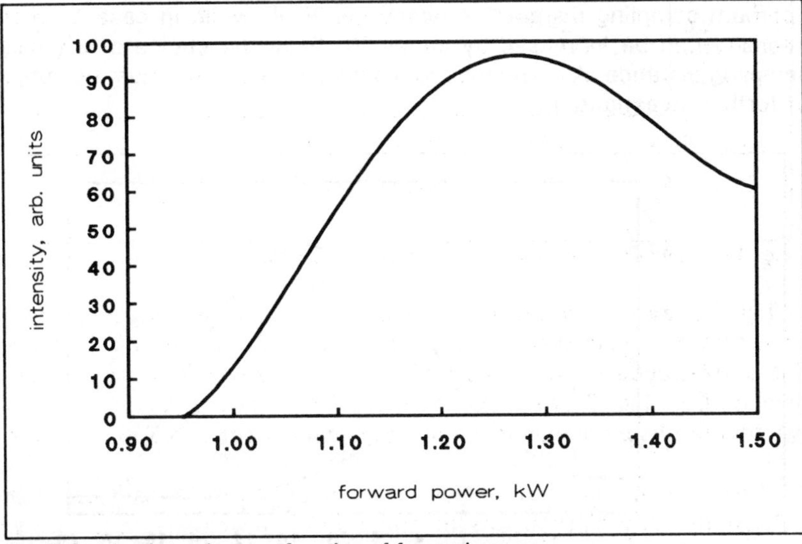

Fig. 5 Cu^+ ion intensity as a function of forward power.

function of the forward power, Fig. 5 shows the result of a measurement with a sampling distance of 11 mm.

For the axial intensity distribution of ions, an additional dependence on the element is observed, which can likewise be explained by the spatial dependence on the ion formation. As a demonstration of this effect, Fig. 6 shows the axial intensity distribution of the elements Li, Cu, Cd and Pb covering the whole mass range. For Cu, Cd and Pb, the

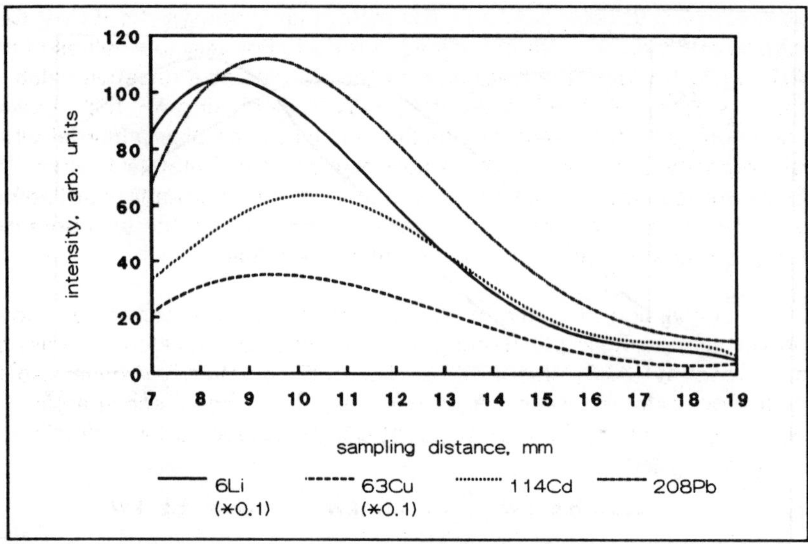

Fig. 6 Axial intensity distribution of some elements.

optimum sampling distance is nearly identical, while in case of Li the intensity can be increased by about 30 % by the choice of a smaller sampling distance. A correlation to a physical parameter must be subject of further investigations.

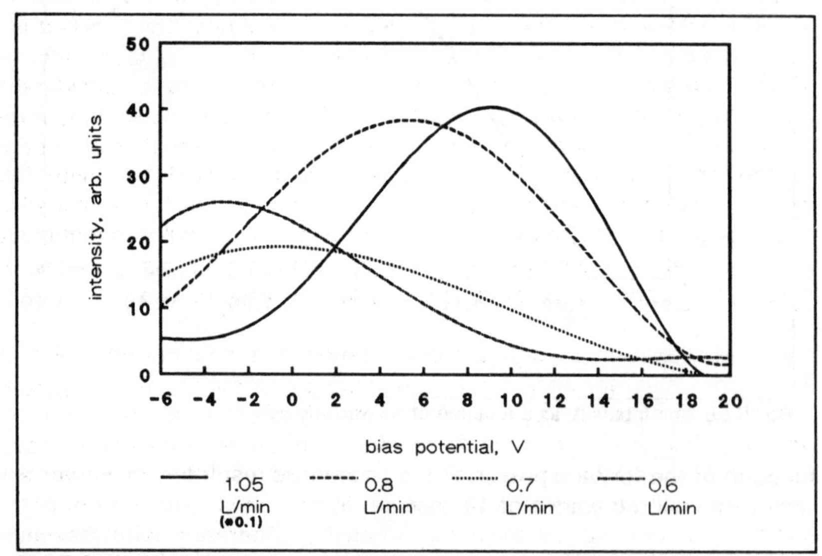

Fig. 7 Cu^+ ion energy characteristic for different aerosol gas flows.

A further parameter of influence in the ICP-MS system used here is the bias potential. The simple optimization of this potential has the same effect as a re-optimization of the lens settings in the operation of commercial instruments. As an example, Fig. 7 shows the IEC of Cu^+ ions as obtained for different values of the AF. For very low values of the AF, e.g. 0.6 L/min, the maximum of the IEC lies at a negative value of the bias potential, in this case at -4 V. With increasing AF, the intensity maximum is shifted towards positive values of the bias potential while the intensity increases. Above a flow rate of approximately 1 L/min, the bias potential does not change any longer. Of course, the forward power is also of influence here. An increase in power shifts the bias potential for the intensity maximum towards negative values.

In conclusion, we have a strong influence by the parameters flow, power and sampling distance as also observed in atomic emission spectroscopy (AES). In ICP-MS unlike AES however, we have also to realize optimum ion transmission with respect to ion kinetic energies for which adjustment of the bias potential was proved to be a simple and effective means.

3 COMBINED OPTIMIZATION

As just discussed, sampling distance, gas flow, power and bias potential exert a strong influence on the ion intensities registered by the MS system. It is common laboratory practice to keep the sampling distance constant and to independently optimize only the standard parameters AF and forward power. This leads to the impression that only a small change in these parameters will result in satisfying intensities which in fact is a self-made restriction of analytical flexibility. Simultaneous optimization in view of the results just presented may be utilized to achieve considerably higher flexibility for analytical applications. In particular the shifting of the axial distribution of ion intensities and of the bias potential as a function of the operation parameters discussed earlier should be taken into account for optimization, because their influence can be exploited in order to considerably extend the useful range of operation parameters. For example, a higher power may be desirable in many cases, in particular for analysis of organic solvents or slurry analysis.

An example of a combined optimization is demonstrated in Fig. 8. The upper diagram shows the maximum ion intensity resulting from optimization of the AF at a defined sampling depth as a function of the power; in the lower part, the AF leading to optimum intensity is shown as a function of the power. By this combined variation of power and flow, the axial position of maximum ion formation is fixed in front of the sampler maintaining the maximum intensity. Operation with maximum power requires an AF of 2.2 L/min. The useful range of the power is now extended to the large interval from 0.8 to 2.1 kW.

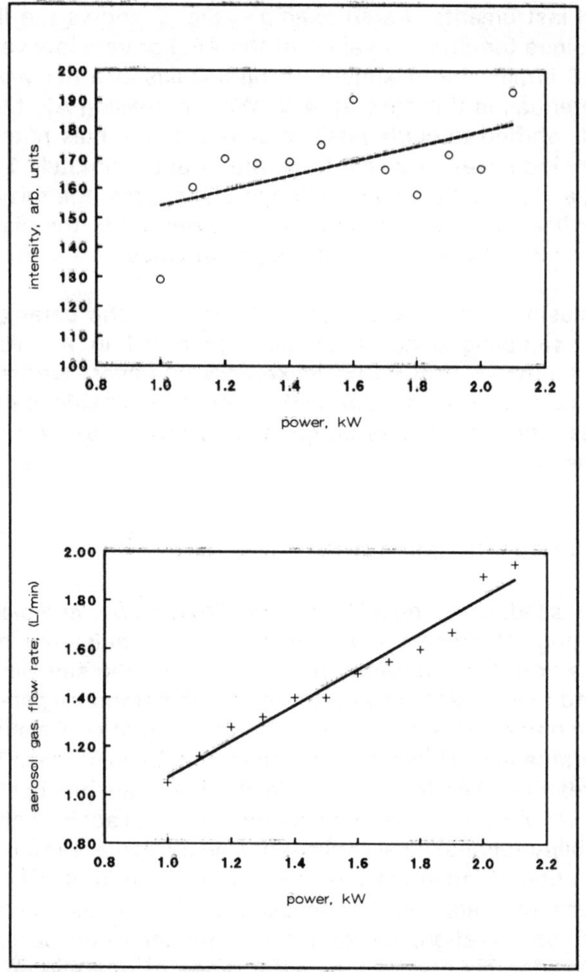

Fig. 8 Combined optimization; upper part: Cu^{+} ion intensity resulting with optimized flow rate, lower part: optimized flow rate.

4 INTERFERING ION SPECIES

Spectral interferences by multiply charged species or molecules impose a considerable restriction in the analytical performance of ICP-MS. Plasma monitoring can give valuable information as to the elucidation of the origin of such interferences. Fig. 9 shows, as an example, the axial distribution of YO^{+}, Y^{2+} and Y^{+} of a 10 ppm Y solution. The maxima of the individual ion species are situated for YO at 8 mm, for Y at 12 mm and for double-charged ions at 16 mm.

Yttrium was chosen because this element has a great bond strength

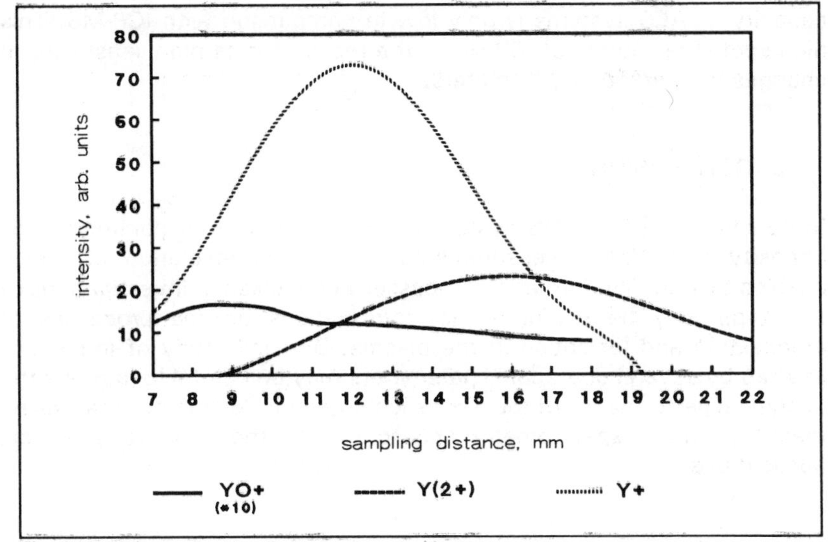

Fig. 9 Axial intensity distribution of some Y species.

to oxides in order to study the oxide formation. The oxide molecule is built directly in the plasma and not in the interface or boundary layer. This can be seen from the red molecular emission line emerging from the plasma. Increase of the single charged ion intensity is correlated with the decrease of the molecular emission intensity, showing that dissociation of the molecule leads to free atoms, which are then ionized. However, a small fraction of these molecules is also registered at the position of the maximum of the analyte ions, so that an interference by oxides may be expected. Reduction of water by cooling of the spray-chamber or desolvation of the wet aerosol can reduce the oxide formation significantly.

The intensity maximum of double-charged ions arises at a greater axial distance than that of single-charged ions. This leads to the conclusion that the formation of multiple-charged ions contributes to the loss of single-charged ions. Thus, variation of the distance enables a certain influence to the ratio of analyte ions and interfering ions.

Support for the results presented here can be found in emission spectroscopic measurements reported by Furuta[9] which show many similarities. In AES the observation height represents an important factor which can be optimized with respect to the actual analytical task. In ICP-MS the sampling depth is the factor of corresponding influence, and it is of similar significance; but in most practical cases, people do not make use of it for optimization.

The measurements by Furuta were performed with very high spatial resolution, which is not standard in AES. Usually, the spatial resolution capacity of AES systems is only low in comparison with ICP-MS. This high spatial resolution of ICP-MS is the reason for its high sensitivity to changes in operational parameters.

5 CONCLUSION

In conclusion, an ICP-MS system can be utilized as a monitor for ion intensity distribution measurements the spatial resolution of which enables to study ion formation processes in the plasma. Investigations of this type may be useful to get informations on the processes of atomization and ionization in the plasma. Detailed study of influences exerted by several operational parameters may be helpful to improve the analytical performance of the ICP as ion source for elemental analysis by means of mass spectrometry and to extend the possible analytical applications.

Acknowledgements - Considerable support by Finnigan MAT, Germany, is gratefully acknowledged. The work was supported by the Bundesminister für Forschung und Technologie and by the Minister für Wissenschaft und Forschung des Landes Nordrhein-Westfalen.

REFERENCES

1. G. Tölg and P. Tschöpel, Anal. Sci., 1987, 3, 199.
2. D.J. Douglas and J.B. French, J. Anal. At. Spectr., 1988, 3, 743.
3. N. Jakubowski, B.J. Raeymaekers, J.A.C. Broekaert and D. Stuewer, Spectrochim. Acta, 1989, 44B, 219.
4. N. Jakubowski, D. Stuewer and W. Vieth, Fresenius Z. Anal. Chem., 1988, 331, 145.
5. G. Peter and K. Hoefler, J. Vac. Sci. Technol., 1987, 5, 2285.
6. G.H. Vickers, D.A. Wilson, G.M. Hieftje, Spectrochim. Acta, 1990, 45B, 499.
7. W.A.H. Van Borm, J.A.C. Broekaert and R. Klockenkämper, Spectrochim. Acta, in press.
8. M.W. Blades, 'Inductively Coupled Plasma Emission Spectroscopy, Part 2', Ed. P.W.J.M. Boumans, Wiley & Sons, New York, 1987.
9. N. Furuta, Spectrochim. Acta, 1986, 41B, 1115.

The Use of ICP-MS to Determine the Retention and Distribution of Platinum in Animals Following the Administration of Cisplatin

P. Tothill[1], K. McKay[2], L.M. Matheson[1], M. Robbins[3], and J.F. Smyth[1]

[1] ICRF MEDICAL ONCOLOGY UNIT, WESTERN GENERAL HOSPITAL, EDINBURGH, UK
[2] SCOTTISH UNIVERSITIES RESEARCH AND REACTOR CENTRE, EAST KILBRIDE, UK
[3] CRC RADIOBIOLOGY RESEARCH GROUP, CHURCHILL HOSPITAL, OXFORD, UK

1 INTRODUCTION

Cis-Diamminedichloroplatinum (II) (CDDP or cisplatin) is used widely in cancer chemotherapy and contributes to the cure of some tumours. Unfortunately there are potentially serious side-effects including sickness, deafness, nerve and blood problems and especially kidney damage. These limit the doses that can be administered and may affect long-term survivors. There is therefore much interest in studying the dynamics of platinum distribution and retention in experimental animals and in patients. Previously used techniques, such as graphite furnace atomic absorption spectrometry, have a limited sensitivity and studies had not extended beyond 2 or 3 weeks. Although it was known that much of the platinum is excreted in the urine in a few days, there had not been a complete acccounting for the administered dose over any period. Access to inductively-coupled plasma mass spectrometry (ICP-MS) facilities offered a large increase in sensitivity and enabled us to plan longer-term experiments.

2 METHODS

Sample preparation

Sample preparation was influenced by the need to minimise the amount of organic matter present and to keep the residual salt concentration below 0.2%. Blood and tissue samples were digested in hot concentrated nitric acid. Amounts up to 1g were treated in test tubes and heated at a nominal 105^{0}C to dryness, a procedure taking a day or more. The residue was dissolved in 1% hydrochloric acid to a 10-fold dilution of the original volume. Bone required dilution by a factor of 200. Larger samples, up to whole rats at 300g, were dissolved in an equal volume of hot nitric acid in a beaker, but not taken to dryness. The non-fat portion was digested in less than 30 minutes. 1% hydrochloric acid was added to give a dilution of about 100g/l and an aliquot filtered off.

Fat was not completely dissolved, but more aggressive digestion of some samples of fat residue showed that their platinum concentration was close to that of the aqueous filtrate, which was therefore considered to be representative of the total digest.

Equipment and technique

The ICP-MS apparatus used was a VG Plasmaquad PQ1, fitted with a Fassel-type torch, a Gilson 222 automatic sample changer and an IBM PC-AT data system. A nebuliser flow rate of 0.7 ml/min was used and a starting volume of 4 ml was required. Indium was used as an internal standard. A portion of a mass spectrum is shown in Figure 1. Initially, complete scans over sections of the spectrum were carried out. Subsequently, peak jumping on the ^{115}In, ^{194}Pt, ^{195}Pt and ^{196}Pt peaks gave a more efficient use of measuring time. Three measurements were made on each sample. The computer established a calibration equation by regression analysis of the results from blanks and standards derived from Specpure platinum.

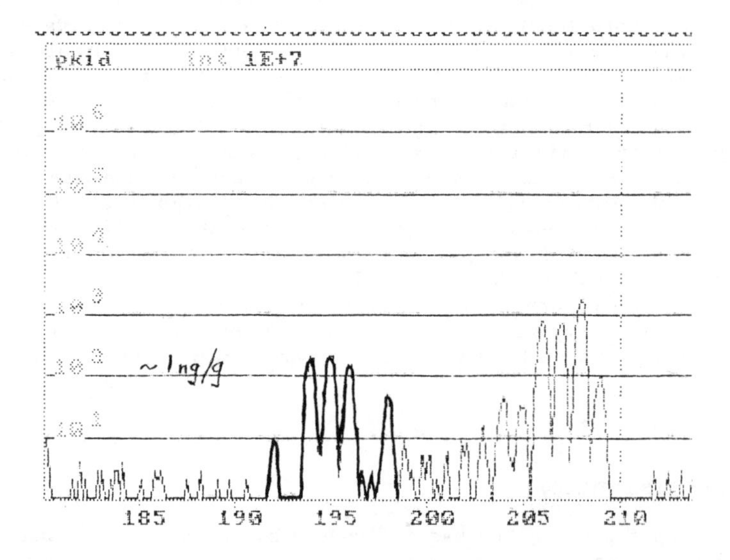

Figure 1 Part of mass spectrum from sample of rat liver, with the platinum isotopes highlighted. Peaks corresponding to lead in the tissue and mercury in the argon gas are also evident.

Reproducibility was good, measurements of standards repeated over a 3 month period being more than 10% outside the nominal value in only 4% of cases. Matrix effects were small, the indium internal standard correcting well for excess salt concentration.

Residual nitric acid had no appreciable effect on sensitivity, but raised the background slightly. The detection limit, defined as the concentration of platinum corresponding to 3 times the standard deviation of repeated measurements of a blank solution, was typically 0.01 ng/ml, giving 0.1ng/g in the original soft tissue.

Further details of the technique have recently been published.[1]

3 ANIMAL EXPERIMENTS

Rats received injections of cisplatin at a dose of 2.5 mg/kg body weight 3 times at weekly intervals, a regime similar to that used for treating patients. They were killed at intervals up to 3 months and measurements made of platinum concentration in most organs and in the whole body. Blood samples were also obtained at intervals. Results were expressed as concentrations in ng Pt/g of tissue and as the percentage of the administered dose in whole organs. Some of the latter results are illustrated in Figure 2.

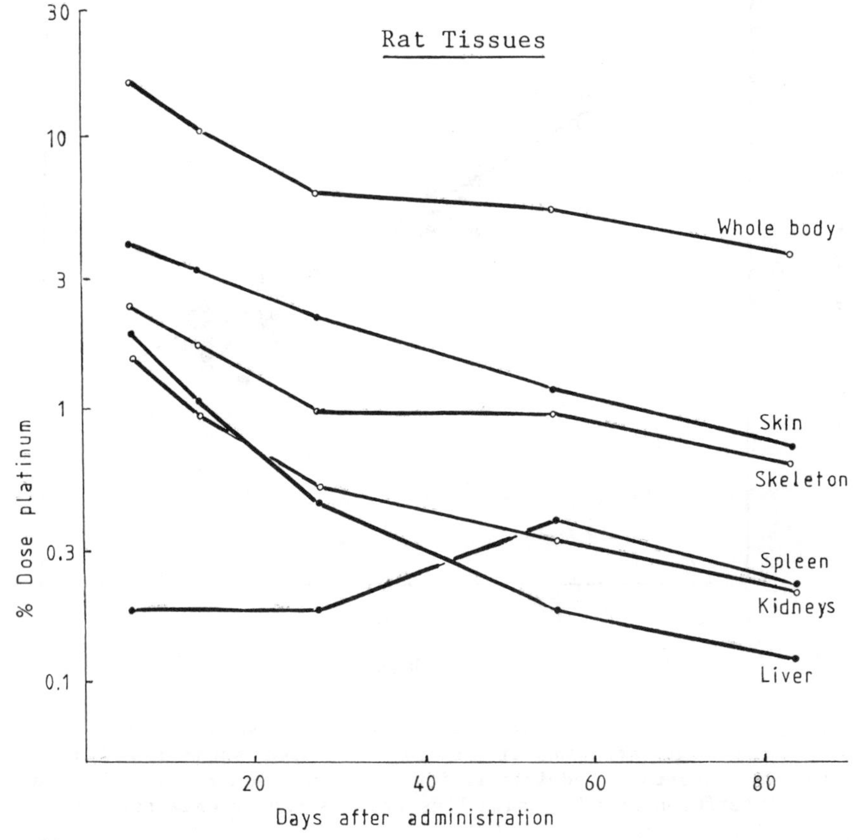

Figure 2 Retention of platinum in whole body and organs of rats at different times after the administration of cisplatin.

The highest concentrations of platinum were in the kidney and spleen, but, because of their greater masses, larger overall quantities were retained in the skin, skeleton and muscle. A mean of about 4% of the dose was still present in the body 3 months after administration. The concentration of platinum in blood plasma and red cells fell more quickly than that in any other tissue (Figure 5).

More limited measurements were made on mice. Nude mice bearing tumour grafts received single injections of cisplatin with or without injections of interferon and killed at intervals up to 40 days. Platinum was measured in the total body and also in the tumours. The results are illustrated in Figure 3, where a log-log scale is used, as retention was found to follow a power function. For comparison, the mean results for rats, plotted in the same way, are included. The slopes are virtually identical, and there was no effect of interferon administration on platinum retention.

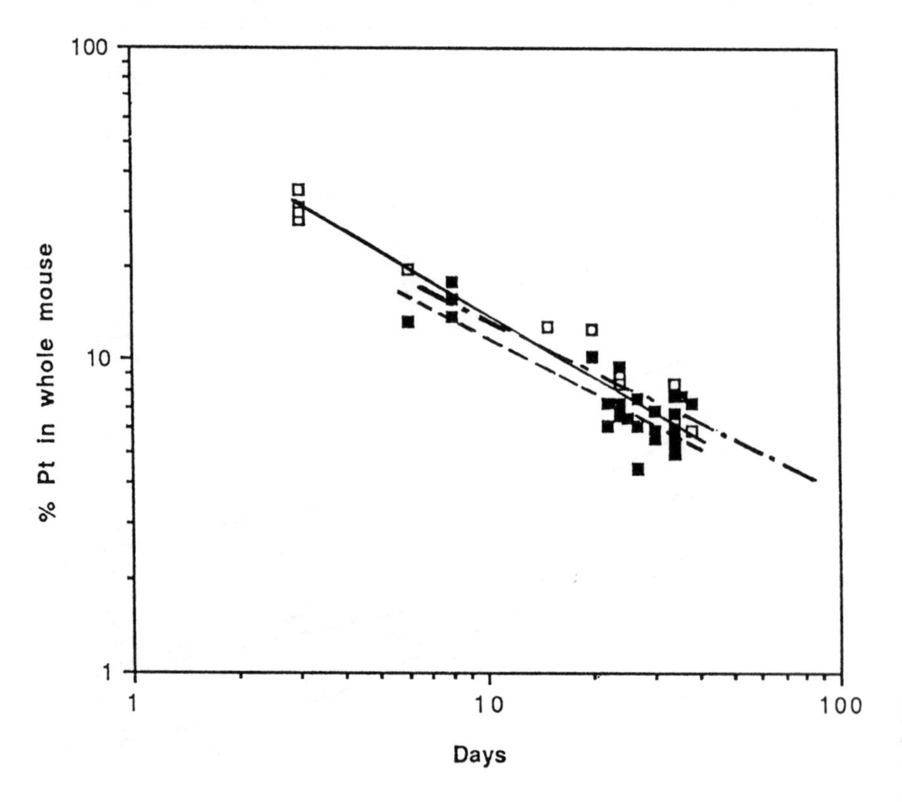

Figure 3 Retention of platinum in nude mice at different times after the administration of cisplatin. Open squares and full line, no interferon administration. Closed squares and broken line, interferon added. Chain line, corresponding data for rats.

Pig kidneys are structurally and functionally closer to those of man than other experimental animals and are therefore being used to study the effect of cisplatin on long-term kidney function damage. This provided the opportunity of obtaining biopsy samples of skin and blood and autopsy samples of most organs over a period of up to one year after cisplatin administration. Some examples of results are plotted in Figure 4. Doses of 1.5, 2.0 and 2.5 mg/kg were infused, so the platinum concentrations have been normalised for dose.

Figure 4 Concentration of platinum is some organs of pigs at different times after the administration of cisplatin.

In contrast to the results for rats, platinum concentration in the liver was higher than that in the kidney and fell more slowly. Elimination of platinum from skin was approximately exponential for about 6 months, but then seemed to level off. We do not have the weights of the major tissues, such as muscle, skin and skeleton, but some reasonable assumptions suggest a retention of about 3% of the administered dose at 1 year.

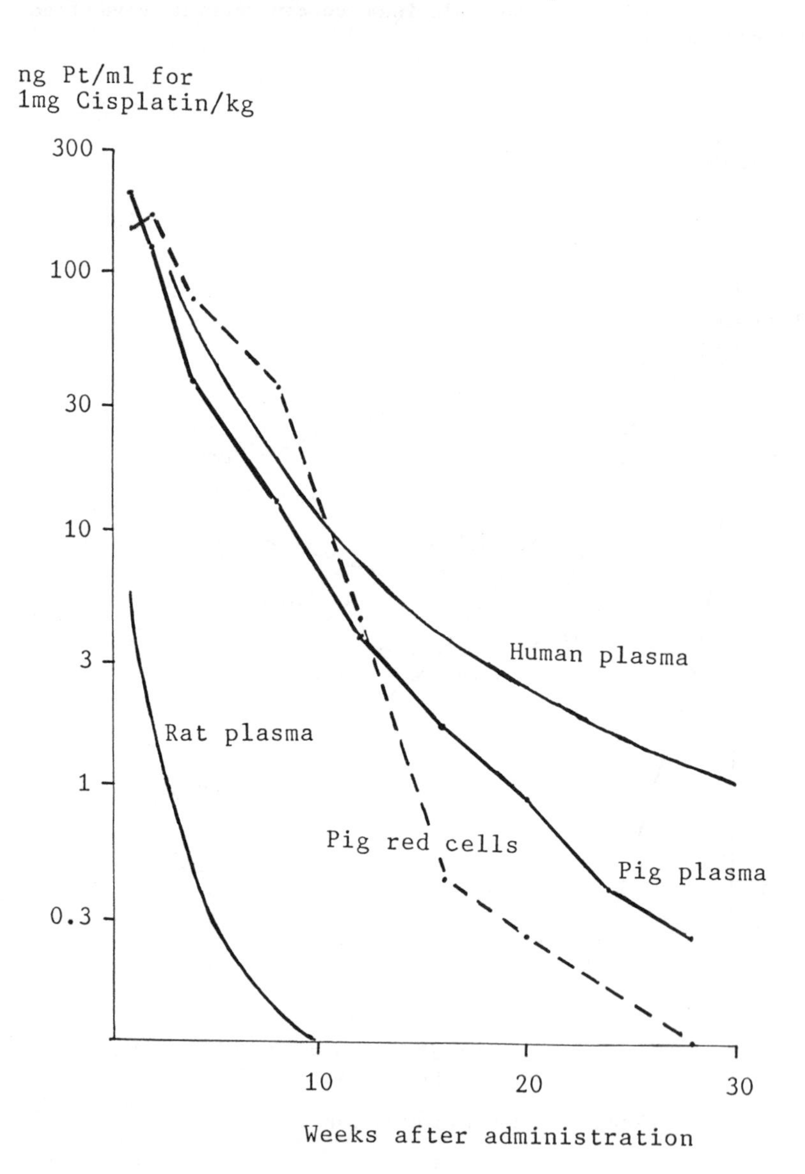

Figure 5 Concentration of platinum in the blood of three species at different times after the administration of cisplatin.

As with rats, platinum concentration in pig blood fell more quickly than that in other tissues, but was slower than the elimination from rat blood, as is seen in Figure 5. Also illustrated are the only data obtained to date from human subjects. Plasma concentration fell in a multi-exponential or power function manner; indeed, we have been able to measure platinum in blood from patients some years after they received their chemotherapy. With pigs the Pt concentration was too low after about 6 months.

4 CONCLUSIONS

The chemical form of the platinum retained in the body is not known and we are not certain of the relevance of our results to long-term morbidity; however, we have established that there are substantial reservoirs of a potentially toxic element retained for an appreciable period. This knowledge may form the basis of further studies, particularly connected with efforts to reduce the toxic side-effects of cisplatin therapy.

REFERENCE

1. P. Tothill, L.M. Matheson, K. McKay and J.F. Smyth, J. Anal.At.Spectrom., 1990, 5, 619.

Whole Blood Analysis by ICP-MS

Th.M. Lutz, P.M.V. Nirel, and B. Schmidt

TA TRACE-ANALYTIC SA, CHEMIN BUVELOT 9, 1110 MORGES, SWITZERLAND

1 INTRODUCTION

The determination of trace element concentrations in human fluids has become a major issue in the field of clinical chemistry. The fast and reliable estimation of element burden is a key factor in many situations . On the upper range of concentrations we have to deal with poisoning and environmental exposure; on the lower range, biomedical research has shown during the recent years, that some trace elements have specific functions in the biochemistry of living organisms. [1]

Analyses of total blood using the classical methods such as Graphite Furnace are slow and cumbersome. ICP-MS represents a powerful tool to determine trace levels of elements. ICP-MS is a fast, sensitive multielementary method, and it necessitates little quantities of sample . All these characteristics represent major advantages in a field where human life can be directly involved.

Most of the impediments in total blood analysis are related to the complexity of the matrix : the high viscosity and the salt-concentration of the blood make a treatment of sample necessary before its injection in most of the analytical devices. The preparation of the sample has to be fast, safe, and unexpensive. Usually the total blood is acid digested. [2] The problem is that this treatment does not fit any of the previously cited requirements.

In this report, we consider only whole blood treatment prior to ICP-MS analysis. However, our experience has shown that our preparation is directly applicable to other clinical samples such as urine, serum or plasma. [3]

2 EXPERIMENTAL

Instrumentation : This work was made on a inductively coupled mass spectrometer, VG PlasmaQuad 2 plus, (VG Elemental, Winsford, Cheshire, UK) under standard conditions.
Samples : All samples used were human Potassium-EDTA whole blood. (Microgen AG, Hofwiesenstr. 370, 8050 Zürich, Switzerland)
Standards made out of 1000 ppm single element solutions. (Bernd Kraft AG, Duisburg, FRG)
Reagents : Water : Milli Q. (Millipore, Bedford, MA, USA) All other reagents were analytical grade. (Flucka AG, Buchs, Switzerland)

SAMPLE PREPARATION

To prepare whole blood for ICP-MS analysis, we have to consider the following problems: concentration of non volatile components, namely NaCl, molecular interferences, viscosity and stickiness, stability of the plasma, stability of the sample, the choice of an internal standard, detection limits, sample volume. Starting from works of Delves et al [4], we optimized the following steps :

Dilution

By simple dilution, the salt concentration, the viscosity and the detection limits fall. The sample volume rises and an internal standard can be added. A dilution factor of 5 is a good compromise between detection limit and salt concentration problems and was found to be an optimum in our preparation method.

Buffering

The goal is to obtain a clear solution of all components. This implies to keep all the proteins intact in their native form. To prevent precipitation, an excellent buffering in the range of pH 8 to 9 is necessary. 5 times diluted potassium EDTA blood has a pH of 6.9 . Addition of ammonia to a concentration of 27 mMol raises the pH to 8.5. At this pH, the solution is clear red and nearly transparent. Addition of more ammonia brings the solution to perfect transparency; it means perfect dissolution. But ammonia has a negative effect on the element response.
We found a compromise between perfect dissolution and good response at a pH of 8.5 i.e 27 mMol ammonia.
In the case of heparin blood, no pH value was found that gives a clear and transparent solution. A second advantage of the EDTA in the blood is that it helps to keep cations in solution.

Addition of a detergent

The addition of a detergent (Triton 100X, 0.1 %) to the sample has different effects. First it helps to prevent blockage of the system and improves the wash out of the sample after the analysis. Secondly the nebulizer efficiency rises. The average droplet size is smaller and brings more sample into the plasma. A higher count rate is observed.

Haemolysis

The most important step in our preparation is the haemolysis. The whole blood is a suspension of 45 volume % of cells in a salt and protein solution. [5] More than 99 % of the cells are erythrocytes, the red blood cells. The other non-dissolved components, namely thromocytes and white blood cells are less important. The high cell concentration is responsible for the characteristic viscosity and stickiness of the blood. The stability of the cells is given by the osmotic equilibrium between cells and solution. In case of blood about 300 mOsmol l^{-1}. By dilution, the osmotic pressure of the solution falls and the cells begin to swell. If the osmotic pressure falls below 100 mOsmol l^{-1}, the erythrocytes break off and the main component, the haemoglobin is liberated and dissolved. The thrombocytes are also dissolved and only membrane fragments remain undissolved. The membranes of the red blood cells are made of polysaccharides which form thin particles of less then 5 micrometers in size after haemolysis.

Chemical interference correction

The high chloride concentration in the whole blood is a big handicap for arsenic determination. Evans et al [6] describe a simple method of interference reduction without separation process. By addition of propan-2-ol to the sample, the formation of argonchloride is strongly suppressed.

The relative response of the internal standard is increased between 0.5 and 2 % of propan-2-ol. This is not due to a positive interference; all elements are showing an increase in the count rate. We believe that this increase is due to a higher nebulizer efficiency. This is related to a decrease of the surface tension of the sample and the formation of smaller droplets. Above 3 % of propan-2-ol the response falls; it can be due to an overload of the plasma and a lower ionization efficiency.

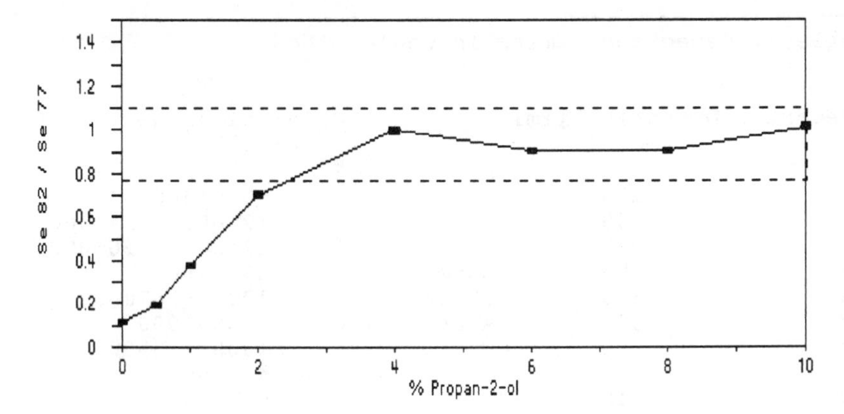

Fig 1 : The efficiency of the interference suppression: The ratio of selenium concentration, calculated with isotopes 82 and 77.

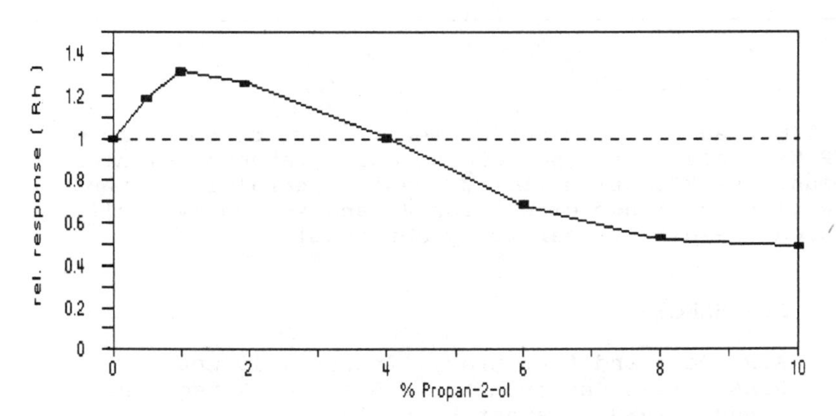

Fig 2 : Response: Relative response of the internal standard (20 ppb Rh)

3 RESULTS

A blood sample prepared as presented is stable at room temperature over weeks. The temperature range for storage is 10 to 30 °C. Below 5 °C and above 50 °C an irreversible precipitation is observed.

To calibrate whole blood analysis, we used an external multielementary standard. The detection limits of a few, clinical important elements in whole blood are shown in Table 1. The clinical range depends on the literature. We used data from Tietz.[7]

Table 1. Detection limits in whole blood.

Element	Detection limit nmol l^{-1}		Clinical range nmol l^{-1}
Mn	100		50 - 500
Cu	50		12'000 - 24'000
Zn	50		12'000 - 20'000
As	300	Normal	30 - 800
As	300	Chronic	1300 - 6700
As	300	Acute toxic	> 800'000
Se	200		1800 - 3500
Cd	20		< 50
Au	10		< 500
Hg	10		< 100
Tl	10		< 40
Pb	10		< 1000
Bi	10		< 50

4 CONCLUSION

Our sample preparation method permits to perform ICP-MS analysis in the concentration range of clinical needs. The detection limits are in general lower then the clinical range except for As and Mn; however all toxical levels are perfectly detectable.

REFERENCES

1. A.R. Date and A.L. Gray, 'Inductively Coupled Plasma Mass Spectrometry', Blackie, Glasgow and London, 1989, Chapter 5, p.115
2. T.D.B. Lyon, G.S. Fell, R.C. Hutton and A.N. Eaton , J. Anal. At. Spectrom.,1988, 3, 265
3. Th.M. Lutz, P.M.V. Nirel, in preparation
4. H.T. Delves and M.J. Campbell, J. Anal. At.Spectrom.,1988, 3, 343
5. W.F.Ganong, 'Physiologie', Springer, Berlin, 1974, Chapter 27, p.435
6. E.H. Evans and L. Ebdon, J. Anal. At. Spectrom.,1989, 4, 299
7. N.W. Tietz, 'Clinical Guide to Laboratory Tests', Saunders, Philadelphia, 1983.

Multivariate Analysis of ICP-Mass Spectra: Determination of Nickel and Iron in Body Fluids

Douglas M. Templeton and Margaret-Anne Vaughan

DEPARTMENT OF CLINICAL BIOCHEMISTRY, UNIVERSITY OF TORONTO, 100 COLLEGE ST., TORONTO, CANADA M5G 1L5

1. INTRODUCTION

A major limitation to the use of ICP-MS in a biomedical context arises because so much interesting biology is inconveniently based on elements of the first transition series. Chromium, Mn, Fe, Co and Cu are all essential elements in man with well defined functions and certain clinical indications for measurement. Nickel is possibly essential [1], and Ni and Cr are potential carcinogens with important sources of industrial exposure [2, 3]. Vanadate is generating much interest as a phosphotyrosine phosphatase inhibitor and promising hypoglycemic agent [4]. If we add Zn to this set of transition elements, we include a major nutrient necessary for cell growth and division, as well as immune function [5]. There is good reason to assume an interest in the widespread measurement of Zn in clinical samples [5], at least for research purposes. Which of these elements can conventional, low resolution ICP-MS handle?

Table 1 is a list of some of the polyatomic interferences known to occur at the masses of the elements V through to Zn. No major isotope of these elements is free from interference. In some cases, the polyatomic species arise from the Ar plasma background. In other cases, elements present at high concentrations in biological materials (e.g. N, Cl, Ca) contribute. Because ICP-MS provides isotopic information, it is particularly appealing to contemplate stable isotope tracer studies in human subjects. However, if one wishes to study the biokinetics of an element, an isotopically enriched source must be distinguished from the endogenous naturally abundant element. For one of the aforementioned set of metals, this requires measurement of a minimum of two of the masses in Table 1, doubling the analyst's problems with interferences.

Partial solutions to overcoming isobaric interferences at the masses of these elements are discussed elsewhere in this volume, including sample separation, solvent removal

Table 1 Polyatomic Interferences for Elements from V to Zn[a]

MASS	ISOTOPES[b]	INTERFERENCES
50	Cr(4.35),V(0.24)	^{36}ArN, ^{34}SO
51	V(99.76)	^{35}ClO, ^{37}ClN
52	Cr(83.76)	^{40}ArC, $^{35}ClOH$, ^{36}SO, ^{36}ArO
53	Cr(9.51)	^{37}ClO
54	Fe(5.82),Cr(2.38)	^{40}ArN, $^{37}ClOH$
55	Mn(100)	$^{40}ArNH$
56	Fe(91.66)	^{40}ArO, ^{40}CaO
57	Fe(2.19)	$^{40}ArOH$, $^{40}CaOH$
58	Ni(67.77),Fe(0.33)	^{42}CaO, NaCl
59	Co(100)	^{43}CaO, $^{42}CaOH$
60	Ni(26.16)	$^{43}CaOH$, ^{44}CaO
61	Ni(1.25)	$^{44}CaOH$
62	Ni(3.66)	^{46}CaO, Na_2O, $Na^{39}K$
63	Cu(69.1)	$^{46}CaOH$, $^{40}ArNa$
64	Zn(48.89),Ni(1.16)	$^{32}SO_2$, $^{32}S_2$, ^{48}CaO
65	Cu(30.9)	$^{33}S^{32}S$, $^{33}SO_2$, $^{48}CaOH$
66	Zn(27.81)	$^{34}SO_2$, $^{34}S^{32}S$, ^{50}VO
67	Zn(4.11)	$^{35}ClO_2$
68	Zn(18.57)	$^{40}ArN_2$, $^{36}SO_2$, ^{52}CrO

[a]From references [6, 7] and the present work.
[b]Natural abundances in parentheses.

and high resolution mass spectrometry. None of these approaches is without its own problems and (or) costs. Therefore, this study was undertaken in order to determine if multivariate analysis could be used to separate mass spectral features arising from individual transition elements, from those due to polyatomic interferences.

Principal components analysis (PCA) is a multivariate technique [8] that relies on data reduction and remodeling to derive the number of components necessary to account adequately for a complex system (e.g. a series of spectra of mixtures of pure components). Its application to chemical spectroscopy has been well described in the monograph by Malinowski and Howery [9]. Because distinct spectral patterns cluster in different regions of data space [10], PCA is well suited to separation of spectral data. A series of factors are derived that in decreasing order of importance describe the information content of the original set of spectra. This gives an indication of the number of significant components in the system, and implicitly the number of interfering components in the spectra. Our implementation of PCA in ICP-MS is based on the use of the IND function [9] and successive eigenratios [10] methods of data reduction, and use of target solutions and the SPOIL function [9] to retrieve concentration data. This has been described in more detail elsewhere [11, 12]. Preliminary studies by PCA of Ni in urine and data presented in Figures 1 and 3 have appeared [11, 12].

2. MATERIALS AND METHODS

Analyses were performed using a Perkin Elmer-Sciex Elan 250 ICP-MS with an extended torch positioned such that the tip of the Cu sampler was 20 mm from the end of the load coil. Samples were nebulized through a Meinhard C nebulizer into a Scott spray chamber. Operating parameters are given in Table 2. A Tylan F2-280 mass flow controller was used to control the inner gas flow rate. Data were acquired at low resolution using the multielement mode. No corrections for isobaric overlap were applied at the time of measurement. Calcium, Ni, Fe, Na, K and Rh standards were prepared from 1000 or 10000 mg/L stock solutions (SPEX Industries Inc.; St. Laurent, Quebec). Enriched ^{62}Ni was obtained from Oak Ridge National Laboratories as Ni powder and was dissolved in Suprapur HNO_3 (Merck; Darmstadt, Germany). [^{57}Fe]- and [^{58}Fe]-ferrous sulphate, gifts of Dr. Stan Zlotkin of The Hospital for Sick Children, Toronto, were originally obtained from Oak Ridge as the corresponding oxides.

Urine samples from healthy volunteers were diluted volumetrically with water to 40% for analysis. Expired human serum from the hospital's blood bank was digested in 30mL PTFE screw-capped vials. Aliquots of 0.5 mL were pipetted into the vials and 0.5 mL of Suprapur HNO_3 was added. The vials were capped and warmed on a hotplate at medium heat for 3 h. Blood samples from volunteers were digested in a similar manner, except that 1 mL of HNO_3 was used for 0.5 mL of whole blood. When cool, the digests were transferred to volumetric flasks and diluted to 5 mL with deionized water. Rhodium was added as an internal standard (0.1 µg/mL) to all samples except the blanks.

Principal components analysis was done on a Macintosh SE/30 computer as described elsewhere [12]. Spectra of pure standards of Ca, Na plus K, or acid digestion blanks were used as target vectors to achieve transformation back into element space of the eigenvectors resulting from factor analysis [11]. Mass sets and target solutions for measurement in the different experimental protocols are given in Table 3.

Table 2 ICP-MS Operating Conditions

R.F. forward power	1.3 kW	Repeats	5
Measurement time	5 s	Points per peak	3
Dwell time	50 ms		
Gas flow rates		Ion lenses	
outer	12 L/min	Bessel box barrel	4.2 V
intermediate	2 L/min	plates	-11.4 V
inner	1 L/min	photon stop	-6.4 V
		Einzel	-16.4 V

Table 3 Mass sets for the determination of Ni and Fe by PCA

	Natural Abundance	Single Tracer	Dual Tracer
Element	Ni	Ni	Fe
Matrix	Serum, Urine	Urine	Whole blood
m/z	{58,60,61,62}	{58,59,60,61,62}	{54,57,58,80,81}
Targets	{Ni,Ca,Na+K}	{Ni,^{62}Ni,Ca,Na+K}	{Fe,^{57}Fe,^{58}Fe, acid blank}

Reproduced data were obtained by summing the product of the target scores and target vectors. Detection limits for the method were estimated for a 99.5% confidence interval, according to the method described by Ketterer et al. [13].

3. RESULTS AND DISCUSSION

Natural Abundance Measurements - Ni in Serum and Urine

From Table 1 it is clear that major interferences with Ni can be expected to arise from oxides and hydroxides of Ca. Nevertheless, the isotopic patterns of naturally abundant Ni and Ca are quite distinct, and so the analysis of Ni in Ca-rich matrices should provide a challenge well suited to a pattern recognition approach. Urine contains high concentrations of Ca, typically up to 200 µg/mL. For these reasons, we analysed urine samples spiked with naturally abundant Ni as an initial test of the method. A two-component PCA using Ni and Ca targets gave reasonable data reproduction at the major Ni isotopes, but an additional minor component(s) was indicated. Therefore, we systematically analysed urea solutions as well as solutions of the major salts found in urine. The Na solution gave a signal at m/z=62, attributable to Na_2O, but of insufficient intensity to account for the signal observed in urine, even taking ^{46}CaO into account. Although K alone gave no signal in this mass range, additional intensity at m/z=62 arose in mixtures of Na and K. This probably represents formation of the ^{23}Na^{39}K dimer, although the other isotopes of K are of too low abundance to observe a K isotope pattern for NaK, and so other effects of K on Na_2O formation cannot be ruled out as a cause. The improvement in data reproduction achieved by a three-component analysis (Ni, Ca and Na+K) is most apparent at m/z=62, as expected (Fig. 1).

To determine if this method is practical for analysis of real samples, we used it to estimate the Ni content of N.I.S.T. Standard Reference Material 2670 "Trace Metals in Urine". Standard addition was used in combination with a three-component PCA to compensate for potential matrix effects. A value of 69 ± 1 ng/mL (±1σ) was determined, in

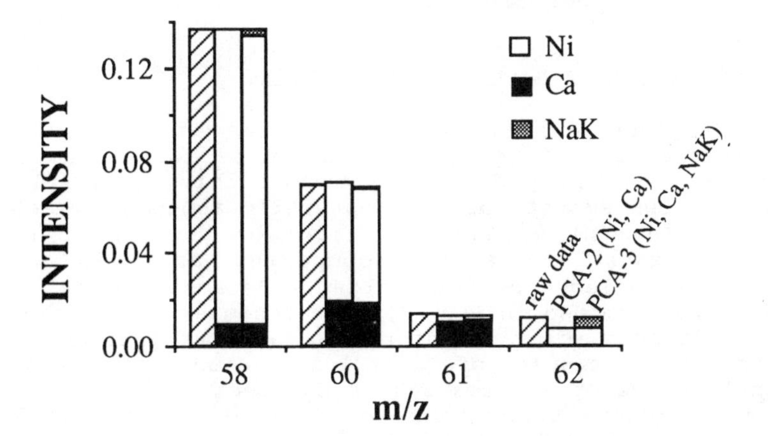

Figure 1 Reproduction of ion intensity data by PCA. Human urine spiked with naturally abundant Ni was analyzed by a two- (middle bars) or three-component (right hand bars) PCA. The ability to reproduce the raw spectral intensities (hatched bars) is compared. The calculated contributions of Ni, Ca and Na+K to the signal are indicated by the appropriate shading. Adapted from [12] with permission.

excellent agreement with the value of 70 ng/mL suggested pending certification. Detection limits were estimated to be 1 ng/mL in 40% urine.

To compare the matrix effects arising from urine and serum, samples of each were spiked with naturally abundant

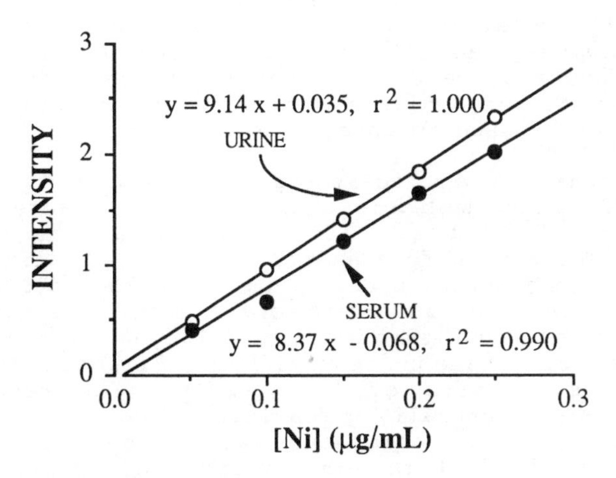

Figure 2 Three-component PCA analysis of human serum and urine spiked with Ni.

Ni and analyzed by PCA (Fig. 2). Additional signal suppression and lower precision are evident in the serum matrix.

Single Isotope Tracer Methods - Ni in Urine

Because isotopically distinct sources of an element produce unique mass spectra, they should be identified as distinct components by PCA. We tested this by spiking urine with known amounts of both naturally abundant Ni and Ni enriched in ^{62}Ni (found 94.85%). Now an additional component exists, so a four-component analysis was undertaken to include the ^{62}Ni target. Because the system must be overdetermined in PCA [9], a fifth mass must now be included that adds relevant information without introducing new components. Scrutiny of Table 1 suggested m/z=59 as an additional source of information on Ca species that has no known interferences other than Co. Because Co generally occurs in urine at levels below the detection limit of ICP-MS, this mass was deemed suitable. A four-component analysis that included m/z=59 in the mass set adequately separated the two isotope sources (Fig. 3).

Dual Isotope Tracer Methods - Fe in Whole Blood

The simultaneous determination of three isotopically distinct sources of an element would allow two different tracers to be measured simultaneously in the presence of the natural element. Such an experiment could find application, for example, in Fe absorption studies. For instance, in healthy adults a large proportion (close to 100%) of newly absorbed Fe in circulation will, after a period of a few weeks, find its way into the erythrocyte. Measurement of erythrocyte Fe is then taken as an indicator of absorption [14]. Factors affecting Fe incorporation, excretion, erythrocyte survival, etc. could invalidate this approach. However, administration of one Fe isotope intravenously would serve as an internal standard for the degree of incorporation, while a second tracer administered orally could then be used to assess absorption.

To determine an enriched isotope against a background of the naturally abundant element, one can measure the ratio of the signals at the enriched mass and at one other isotope of the element, before and after use of the tracer, and then derive concentration data from an isotope dilution equation. When two tracers are used, at least two ratios must be determined, and the isotope dilution expression extends to two simultaneous equations. Propogation of errors during evaluation of the resulting expression renders the approach generally unfeasible. However, PCA is able to handle multiple components well, the practical limitation occurring when some individual component patterns are insufficiently different from each other rel-

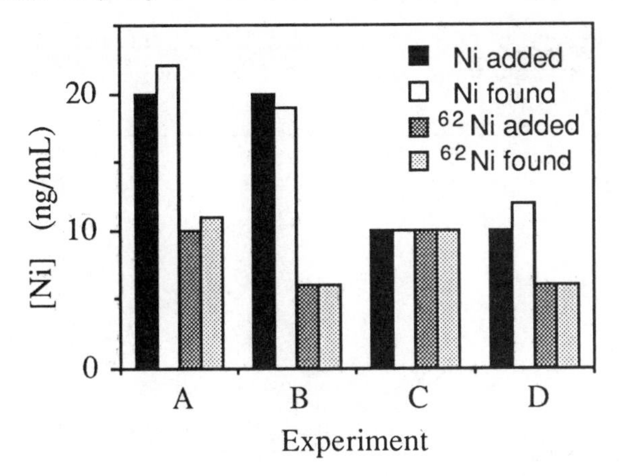

Figure 3 Determination by four-component PCA of Ni in urine spiked with the metal from two isotopically distinct sources. The graph is constructed from data in [12].

ative to a dominant component. In general, isotope enrichment is of a sufficient degree to avoid this problem and sources can be adequately distinguished. Accordingly, we evaluated the ability of PCA to separate sources of Fe enriched with ^{57}Fe and ^{58}Fe, in the presence of naturally abundant Fe.

At the levels of Fe encountered in whole blood, the only interferences expected to be significant are from Ar species. Therefore we used an acid digestion blank as a

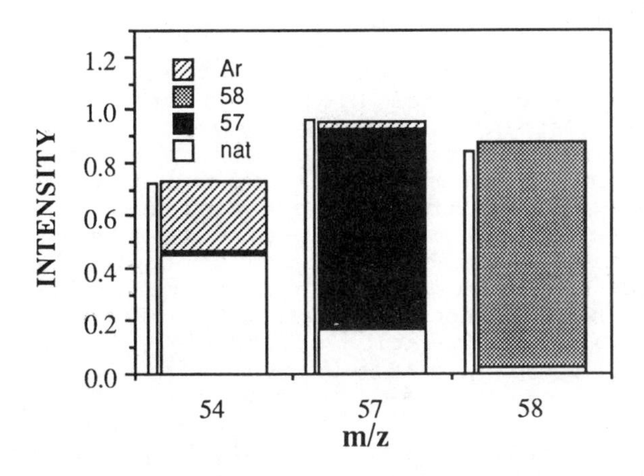

Figure 4 Reproduction of the raw data after four-component PCA of a whole blood digest spiked with ^{57}Fe and ^{58}Fe, showing the proportions of these components as well as naturally abundant Fe (nat) and presumed Ar species interferents. The slim bar to the left of each mass plot indicates the corresponding intensity of the raw data.

Table 4 Four-component PCA of a whole blood digest diluted 1:50 and spiked with enriched ^{57}Fe and ^{58}Fe.

Fe added (μg/mL)		Fe measured (μg/mL)		
as ^{57}Fe source	as ^{58}Fe source	as naturally abundant Fe	as ^{57}Fe source	as ^{58}Fe source
-	-	7.7	0.00	0.00
0.50	-	7.7	0.49	0.00
1.00	-	7.6	1.00	0.01
-	0.50	7.7	0.00	0.47
-	1.00	7.7	0.00	0.98
1.00	1.00	7.6	0.99	1.03

target vector in a four-component PCA to represent the background of the plasma. The other targets were the three Fe solutions. Masses 80 and 81 (representing $^{40}Ar_2$ and $^{40}Ar_2H$) were included in the data set to provide information on Ar species without introducing additional significant components. A digest of whole blood was diluted 1:50 with deionized water to give a solution of 7.7 μg/mL of Fe, as determined independently by atomic absorption. This solution was spiked with 1.0 μg/mL of Fe from both ^{57}Fe- and ^{58}Fe-enriched sources. A three-component PCA that considered only endogenous and added Fe could not account for the observed signal intensity and gave quite erroneous proportions of the two enriched sources. Inclusion of the blank target as a fourth component adequately modeled the system (Fig. 4). Data obtained at different levels of spiking with the two sources (Table 4) demonstrate acceptable determinations by this method.

4. SUMMARY

These results demonstrate that principal components analysis can separate Fe and Ni spectra from important interferents in selected biological matrices. The general utility of the approach remains to be investigated, and other multivariate algorithms should be evaluated. It may be anticipated that problems will arise in choosing mass sets that overdetermine many systems, and the minimal pattern information in spectra of mono-isotopic elements such as Mn and Co renders the approach impractical for these metals. However, when applicable, PCA offers several advantages. It allows separation of interferences without the need for sample manipulation or instrument modification. It can be used to separate isotopically distinct sources of an element as long as they differ sufficiently in spectral pattern, and this does not require high purity of the tracer(s). This may be an important consideration because the cost of stable isotopes escalates with the degree of enrichment. Finally, it facilitates experiments with multiple tracers of the same element which would be difficult by conventional isotope dilution.

Acknowledgements: This work was supported by grants from the Province of Ontario (U.R.I.F.) and the Ni Producers Environmental Research Association (NiPERA).

5. REFERENCES

1. F. H. Nielsen, Annu. Rev. Nutr., 1984, 4, 21.
2. R. F. Hertel, in 'Environmental Carcinogens: Selected Methods of Analysis, Vol. 8 - Some Metals', (I. K. O'Neill, P. Schuller and L. Fishbein, eds.), IARC, Lyons, 1985. p. 63
3. F. W. Sunderman Jr, in 'Environmental Carcinogens: Selected Methods of Analysis, Vol. 8 - Some Metals', (I. K. O'Neill, P. Schuller and L. Fishbein, eds.), IARC, Lyons, 1985. p.79
4. R. I. Dorn and M. J. DeNiro, Science, 1985, 227, 1474.
5. C. F. Mills, ed. 'Zinc in Human Biology', Springer-Verlag, London, 1989.
6. S. H. Tan and G. Horlick, Appl. Spectrosc., 1986, 40, 445.
7. M. A. Vaughan and G. Horlick, Appl. Spectrosc., 1986, 40, 434.
8. K. R. Beebe and K. R. Kowalski, Anal. Chem., 1987, 59, 1007.
9. E. R. Malinowski and D. G. Howery, 'Factor Analysis in Chemistry', John Wiley and Sons, New York, N.Y., 1980.
10. D. F. Wirsz and M. W. Blades, Anal. Chem., 1986, 58, 51.
11. D. M. Templeton and M. A. Vaughan, in 'Trace Elements in Health and Disease', (A. Aitio, ed.), Royal Society of Chemistry, Cambridge, U.K., 1991, p.19 .
12. M. A. Vaughan and D. M. Templeton, Appl. Spectrosc., 1990, 10, (In Press).
13. M. E. Ketterer and J. J. Reschl, Anal. Chem., 1989, 61, 2031.
14. M. Janghorbani, B. T. G. Ting and S. J. Fomon, Amer. J. Hematol., 1986, 21, 277.

Microwave Dissolution and ICP-MS for the Determination of Trace and Ultra-trace Elements in Plant Tissue

S. Yamasaki[1], A. Tsumura[1], and D. Cai[2]

[1] NATIONAL INSTITUTE OF AGRO-ENVIRONMENTAL SCIENCES, 3-1-1 KANNONDAI, TSUKUBA, IBARAKI, JAPAN 305
[2] CHINA HENAN ACADEMY OF SCIENCE, GEOGRAPHY RESEARCH INSTITUTE, ZHENGZHOU CITY, HENAN PROVINCE, CHINA

1 INTRODUCTION

Although the use of $HClO_4$ and/or H_2SO_4 does not seem to be advisable in ICP-MS because these acids are known to produce various molecular species which overlap with several important isotopes[1], digesting plant materials using only HNO_3 by the conventional hot plate and other thermal techniques results in incomplete decomposition of the organic matter in most cases and, therefore, may lead to a poor recovery of the elements. The recent introduction of microwave dissolution technique appears to be a promising method for overcoming the above problems[2-3].

Various standard reference samples of plant materials were digested by the above technique using only HNO_3, and the results obtained by the ICP-MS measurements were compared with the recommended and/or reported values to fully evaluate the suitability of this technique to be used as a sample decomposition method for ICP-MS.

2 EXPERIMENTAL SECTION

Instrumental

ICP-MS. The VG PlasmaQuad (VG Elemental, Ltd., Winsford, Cheshire, England) was used for the most part of these experiments. But some parts of the experiments were also carried out by using a specially designated double focusing type of ICP-MS (PlasmaTrace) with much higher resolution, also provided by the same manufacturer. Although this instrument has been newly developed to resolve molecular overlaps due to matrix species, it has become evident that the detection limits for most elements are significantly improved when the instrument is operated in the low resolution mode[4]. Operating conditions and analytical parameters were essentially those recommended by the manufacturer.

Ultrasonic Nebulizer. The ultrasonic nebulizer used

in this work was provided by Applied Research Laboratories (En Vallaire, CH-1024, Ecublens, Switzerland). A peristaltic pump is used to deliver sample solution to the oscillating surface of the quartz plate. The temperature of the heated tube and the cooling bath was respectively kept at 120 C and 1 C. The flow rate of thesample solution was 2 - 4 ml/min.

Microwave Oven. A microwave oven developed for use in the laboratory, Model MDS-81, purchased from CEM Corp., (Indian Trail, North Carolina, USA.) was used in this experiment. The oven has a power range of 0 to 100% (600 W) with 1% increments and a variable time range of up to 99 hours. The digestion vessel consists of containers of 100 ml capacity, caps (lids) and relief valve. These are all made of Teflon and were also supplied by the manufacturer. A special tray, called a carousel, is also provided and it is possible to load 12 containers in it.

Standard Reference Materials

The reference materials used in this experiment were Tea Leaves, Rice Flour (a), Rice Flour (b), Rice Flour (c), Pepperbush, Chlorella and Sargasso provided by the National Institute for Environmental Studies (Japan), and Citrus Leaves, Tomato Leaves and Pine Needles provided by the National Institute of Standards and Technology (National Bureau of Standards). More detailed descriptions of these materials have been presented elsewhere[5-12].

Reagents

Ultra-high purity nitric acid (Tamapure-100) provided from Tama Chemical Industry Co., Ltd., (Tokyo, Japan) was used. The contents of various metals are guaranteed less than 100 ppt. The water used was purified by double distillation of deionized water using a distiller made of high purity quartz glassware.

Sample Treatment

Half a gram of the sample was weighed directly into a 100 ml digestion container and 10 ml of 70% HNO_3 was added. The containers were sonificated for 2 min to ensure complete dispersion of the sample into HNO_3. The container lids were loosely closed without using the relief valve and stood overnight at about 80 C. This allows the gaseous materials which evolve during the pretreatment to escape from the container, and prevents an increase of the pressure during the digestion at higher temperature. The relief valve was set and the lid was tightly closed. The sealed containers were then positioned in the microwave carousel. At first, the system was operated at full power for 2 min. If the leakage of gaseous materials was not observed, then the digestion was continued for 2 hours at full power.

After cooling, the digest was quantitatively transferred to a 100 ml volumetric flask. One ml of 10 ppm In solution and 1 ml of 1 ppm Rh solution were added and finally made up to 100 ml. When undissolved materials were observed (SiO_2), they were filtered off. The first 20 - 30 ml of the filtrate was discarded and the rest was used for further analysis.

3 RESULTS AND DISCUSSION

Elements of Relatively High Concentrations

The analytical results of Cu and Zn are summarized in Table 1 together with the certified values, and those of Rb and Sr are listed in Table 2. These 4 elements are selected because their contents in the provided standard reference materials are relatively high, and therefore, certified values are obtained with the exception of Zn in Pine Needle, Rb in Tea Leaves and Chlorella, and Sr in Tea Leaves.

These 4 elements, as well as several other elements, were determined by the scanning mode using In (100 ppb) as an internal standard. The time needed for the measurements were around 1 min for each sample. Our results agreed well with the certified values for Cu and Zn. With the exception of Zn in Rice Flour, our values were all within the allowable limits (not shown in Table 1 and 2) of the certified values.

Agreement was also excellent for Rb and Sr, though our values for Tomato Leaves were smaller than the lower limits of the certified values both for Rb and Sr. The discrepancy was very much noticeable for Sr. In the case of Sargasso, it was not possible to obtain a reliable value for Sr, because the concentration of Sr in this sample was too high.

Table 1 Copper and Zinc in Concentrations Standard Reference Materials

Sample	Cu (ppm)		Zn (ppm)	
	Cert.	Found	Cert.	Found
Tea Leaves	7.0	7.0	33	32
Rice Flour-a	3.5	3.6	25.2	23.0
Rice Flour-b	3.3	3.4	22.3	21.3
Rice Flour-c	4.1	4.5	23.1	22.6
Pepperbush	12	12	340	342
Chlorella	3.5	3.3	20.5	20.2
Sargasso	4.9	5.1	16.4	17.6
Citrus Leaves	16.5	16.8	29	29
Tomato Leaves	11	11	62	63
Pine Needle	3.0	3.0		67

Table 2 Rubidium and Strontium Concentrations in Standard Reference Materials

Sample	Rb (ppm)		Sr (ppm)	
	Cert.	Found	Cert.	Found
Tea Leaves		6.1	3.7*	3.5
Rice Flour-a	4.5	4.2	0.3	0.4
Rice Flour-b	3.3	3.1	0.3	0.4
Rice Flour-c	5.7	5.4	0.2	0.3
Pepperbush	75	74	36	34
Chlorella		1.6	40	35
Sargasso	24	24	1010	
Citrus Leaves	4.84	4.86	100	101
Tomato Leaves	16.5	15.2	44.9	39.3
Pine Needle	11.7	11.2	4.8	4.2

* Non certified values

Elements of Low Concentrations

Cadmium was chosen as one of the elements belonging to this category. Only poor agreement was observed for Tea Leaves, Rice Flour-a, Rice Flour-b, and Sargasso. This is simply because the concentration of Cd in these samples is so low that the integrated counts of Cd were only slightly higher than those of the blank solutions. The most simple and straightforward countermeasure in overcoming this problem will be to increase the number of scanning so as to accumulate more total counts. As is also shown in Table 3, analytical results thus obtained are very much improved and coincided well with the

Table 3 Cadmium in Standard Reference Materials (ppm)

Sample	Found		Certified Value
	Method A[a]	Method B[b]	
Tea Leaves	0.11	0.03	0.030 ± 0.003
Rice Flour-a	0.11	0.02	0.023 ± 0.003
Rice Flour-b	0.33	0.31	0.32 ± 0.02
Rice Flour-c	1.82	1.88	1.82 ± 0.06
Pepperbush	6.77	6.69	6.7 ± 0.5
Chlorella	0.09	0.03	(0.026)*
Sargasso	0.14	0.16	0.15 ± 0.02
Citrus Leaves	0.07	0.05	0.03 ± 0.01
Tomato Leaves	2.82	2.70	(3)*
Pine Needle	0.21	0.23	(<0.5)*

* Non certified values
Number of scannings is 120[a] and 960[b]

Table 4 Cobalt in Standard Reference Materials (ppm)

Sample	Found		(Non-) Certified Value
	QP*	HR**	
Tea Leaves	0.155	0.160	(0.12)
Rice Flour-a	0.063	0.027	(0.02)
Rice Flour-b	0.063	0.025	(0.02)
Rice Flour-c	0.053	0.040	(0.007)
Pepperbush	20.4		23 ± 3
Chlorella	0.81	1.03	0.87± 0.05
Sargasso	0.025	0.19	0.12± 0.01
Citrus Leaves		0.145	(0.02)
Tomato Leaves	0.38	0.71	(0.60)
Pine Needle	0.21	0.165	(0.11)

*Quadrupole; ** High Resolution.

certified or recommended values. But because it requires a much longer time for the measurements, this approach will not be regarded as appropriate in cases where analytical efficiency is a major concern.

Use of High Resolution ICP-MS. It is considered possible to solve the above problems if a high resolution instrument is used rather than a quadrupole type ICP-MS because the background level in the former is reported to be about 2 orders of magnitude lower than the latter.

Table 5 Cesium in Standard Reference Materials (ppm)

Sample	Found		(Non-) Certified Value
	QP*	HR**	
Tea Leaves	0.68	0.44	(0.022)
Rice Flour-a	0.45	0.20	(0.70)
Rice Flour-b	0.56	0.33	(0.22)
Rice Flour-c	0.33	0.31	(0.08)
Pepperbush	0.84	1.13	(1.30)
Chlorella	1.22	1.12	
Sargasso	0.17	0.30	(0.04)
Citrus Leaves	1.10	1.34	0.8 ± 0.2
Tomato Leaves			4.5 ± 0.5
Pine Needle	2.61		2.6 ± 0.2

*Quadrupole; ** High Resolution.

Table 6 Lead in Standard Reference Materials (ppm)

Sample	Found		(Non-) Certified Value
	QP*	HR**	
Tea Leaves	0.68	0.91	0.80± 0.03
Rice Flour-a	1.00	1.11	
Rice Flour-b	1.27	1.28	
Rice Flour-c	0.61	0.71	
Pepperbush	6.18		
Chlorella	0.43	0.61	(0.60)
Sargasso	0.89	1.37	1.35± 0.05
Citrus Leaves	11.6		13.3 ± 2.4
Tomato Leaves	6.1		6.3 ± 0.3
Pine Needle	10.4		10.8 ± 0.5

*Quadrupole; ** High Resolution.

Substantial improvements were observed for such elements as Co, Cs and Pb as were presented in Table 4, 5 and 6 when the concentrations of these elements in the reference material are lower than 0.5 - 0.1 ppm. Although the time needed for the measurements using the high resolution ICP-MS is somewhat longer than that of the quadrupole ICP-MS to obtain nearly equal integrated counts, it can be concluded that the use of a high resolution ICP-MS is quite effective for those elements that are contained less than 0.5 ppm in the samples.

Use of Ultrasonic Nebulizer. The use of an ultrasonic nebulizer instead of the commonly used pneumatic nebulizer to improve the efficiency of sample introduction into the plasma will be another possibility for overcoming those problems associated with elements of low concentrations. The analytical results of Mo obtained by the quadrupole ICP-MS combined with an ultrasonic nebulizer are listed in Table 7, together with those obtained by the high resolution ICP-MS. The values obtained by the two above mentioned methods are nearly identical, and are much better than those obtained by the quadrupole ICP-MS alone, with the exception of Pepperbush which contains relatively high amounts of Mo as compared with other reference materials.

A similar approach was also applied for several other elements of very low concentrations. Successful results were, however, not obtained for all the elements examined. These disappointing results are due to much more severe matrix effects commonly observed when sample solutions were introduced through the ultrasonic nebulizer because the amount of solute which reached the plasma was considerably increased. The excellent results observed only for Mo might be due to the fact that the mass number of the internal standard element (Rh) employed in this experiment is relatively close to that of Mo.

Table 7 Molybdenum in Standard Reference Materials (ppm)

Sample	Found			Certified Value
	QP[1]	QP+USN[2]	HR[3]	
Tea Leaves		0.03[5]	0.02	
Rice Flour-a	0.3	0.37	0.37	0.35± 0.05
Rice Flour-b	0.33	0.47	0.41	0.42± 0.05
Rice Flour-c	1.48	1.67	1.60	1.60± 0.1
Pepperbush	0.38	0.53	0.53	
Chlorella	0.88	1.10	1.10	
Sargasso	0.04[3]	0.18	0.20	
Citrus Leaves		0.13	0.14	0.17± 0.09
Tomato Leaves	0.30	0.53	0.50	
Pine Needle		0.10	0.06	

[1] Quadrupole; [2] Ultrasonic Nebulizer ; [3] High Resolution.

Accordingly, it seems it is possible to obtain better results if certain appropriate internal standards are chosen for each element. This possibility was, however, not examined because the procedure for the sample solution treatment as well as concentration measurements became too complicated.

Some Problematic Elements

In spite of its relatively high concentration in the reference materials, the values obtained in this work are significantly higher than the certified values in Cr even when the high resolution ICP-MS is used, as is shown in Table 7. This tendency was more noticeable when the concentration of Cr in the samples was less than 1 ppm.

It seems most probable that these big discrepancies observed in Cr are due to the interference of ^{40}ArC on ^{52}Cr. Although decomposition of the samples by the proposed method appeared to be complete from the color and turbidity of the acid digests, a small amount of undecomposed organic matter must still remain in the solution. The carbon derived from this organic matter is considered to be the main source of the above ArO. Because the resolution needed to separate the peak of Cr from that of ArC is only around 2500, it is well within the ability of the high resolution ICP-MS. However, considerable decreases of the signal intensities are rather common when the instrument is operated in the high resolution mode. Accordingly, it seems most unlikely that it is possible to solve this problem by the above approach unless the concentration of Cr in the sample is relatively high.

Table 8 Chromium in Standard Reference Materials (ppm)

Sample	Found		(Non-) Certified Value
	QP*	HR**	
Tea Leaves	0.68	0.44	(0.15)
Rice Flour-a	0.45	0.20	(0.70)
Rice Flour-b	0.56	0.33	(0.22)
Rice Flour-c	0.33	0.31	(0.08)
Pepperbush	0.84	1.13	(1.30)
Chlorella	1.22	1.12	
Sargasso	0.17	0.30	(0.2)
Citrus Leaves	1.10	1.34	0.8 ± 0.2
Tomato Leaves	3.76		4.5 ± 0.5
Pine Needle	2.61		2.6 ± 0.2

*Quadrupole; ** High Resolution.

Somewhat similar results were also obtained for Ni, though the magnitude and the direction of the deviations from the certified values were not so consistent as those of Cr. Possible causes of these poor results are considered to be spectral interferences of $^{23}Na^{35}Cl$ and $^{42}Ca^{16}O$ as well as the contamination from the skimmer and sampling cones. Judging from the contents of major elements in the reference materials used in this experiment, the interference of NaCl was more probable as the cause of the higher values of Ni in this work.

Table 9 Nickel in Standard Reference Materials (ppm)

Sample	Found		(Non-) Certified Value
	QP*	HR**	
Tea Leaves	8.25	8.72	6.5 ± 0.3
Rice Flour-a	3.76	0.38	0.19± 0.03
Rice Flour-b	3.04	0.54	0.39± 0.04
Rice Flour-c	1.70	0.45	0.30± 0.03
Pepperbush	9.17	10.7	8.7 ± 0.6
Chlorella	12.53	13.4	
Sargasso	3.37	2.04	
Citrus Leaves	0.30	1.66	0.6 ± 0.3
Tomato Leaves	2.93	7.04	
Pine Needle	0.60	4.49	(3.5)

*Quadrupole; ** High Resolution.

4 SUMMARY

The microwave dissolution technique was fully examined in order to establish a sample decomposition method without using $HClO_4$ and/or H_2SO_4, which are known to produce various molecular species that overlap with several important isotopes in ICP-MS analyses. It was possible to decompose most of the plant materials nearly completely by HNO_3 alone. Concentration values obtained by the proposed method were in good agreement with the certified and/or the reference values for all the elements examined, with the exception of Cr and Ni. Although the amount of decomposable samples was around 0.5 g maximum, it was found possible to overcome this disadvantage when the number of scannings was increased or a double focusing instrument was used.

REFERENCES

1. M. A. Vaughan and G. Horlick, Appl. Spectrosc., 1986, 40, 434.
2. S. A. Borman, Anal. Chem., 1986, 58, 1424A; 1988, 60, 715A
3. H. M. Kingston and L. B. Jassie, 'Introduction to Microwave Sample Preparation: Theory and Practice', American Chemical Society, Washington, 1988.
4. N. Bradshaw, E. F. H. Hall and N. E. Sanderson, J. Anal. At. Spectrom., 1989, 4, 801
5. K. Okamoto and K. Fuwa, Environmental Research Quarterly, 1986, 62, 167.
6. K. Okamoto, Environmental Research Quarterly, 1989, 74, 101.
7. K. Okamoto (ed.), 'Preparation, Analysis and Certification of PEPPERBUSH Standard Reference Material', Research Report from the National Institute for Environmental Studies, 18. 1980
8. K. Okamoto and K. Fuwa, Environmental Research Quarterly, 1983, 42, 114.
9. K. Okamoto, Environmental Research Quarterly, 1988, 70, 130.
10. National Bureau of Standards, 'Certificate of Analysis, Standard Reference Material 1572.' 1982.
11. National Bureau of Standards, 'Certificate of Analysis, Standard Reference Material 1573.' 1976.
12. National Bureau of Standards, 'Certificate of Analysis, Standard Reference Material 1575.' 1976.

Determination of Ultra-trace Levels of Rare Earth Elements in Terrestrial Water by High Resolution ICP-MS with an Ultrasonic Nebulizer

A. Tsumura and S. Yamasaki

NATIONAL INSTITUTE OF AGRO-ENVIRONMENTAL SCIENCES, 3-1-1 KANNONDAI, TSUKUBA, IBARAKI, JAPAN 305

1 INTRODUCTION

The contents of rare earth elements (REEs) in a wide variety of environmental samples, especially those of the lanthanides (La-Lu), have been receiving increased attention in recent years because these elements possess exceptionally similar physical as well as chemical properties, and therefore, can be used for tracers of various processes which occur in the environment. In addition to this, REEs are increasingly consumed in the modern industries for the production of numerous new materials. There is a strong possibility that these industrial products will be disposed in the environment. It is, therefore, very important to clarify the background levels of REEs in water samples while the environment is still not contaminated by human activity.

Although it was found that it was possible to determine REEs and several other elements in water samples by ICP-MS after coprecipitation with $Fe(OH)_3$[1], the procedure is still too time-consuming to be used for a large number of samples.

The main purpose of the development of the high resolution ICP-MS is to resolve molecular overlaps due to matrix species[2-3]. However, it has also become evident that the detection limits of most elements are significantly improved when the instrument is operated in the low resolution mode because of the much lower background signals. This is because the long and curved ion flight path together with the complex focusing elements prevent the incidence of photons to the detector[2].

Meanwhile, it has been pointed out that the use of ultrasonic nebulizers significantly improves the sample introduction efficiency [4-6] and has resulted in much increased sensitivity in ICP-AES. We have attempted to combine the above two instruments, and have examined their capabilities by analyzing ultra-trace levels of REEs in water samples.

2 EXPERIMENTAL SECTION

Instrumental

High Resolution ICP-MS. The measurements were carried out using a double focusing type of ICP-MS with much higher resolution supplied by VG Elemental, Winsford, Cheshire, England. The system is illustrated schematically in Figure 1. The ICP torch box and the interface region are basically similar to those used in quadrupole type instruments. The whole interface system is electrically isolated and high voltage is applied to the sampling and skimmer cones. The double focusing mass spectrometer is consist of a 70 electrostatic sector for energy focusing and a 35 laminated magnetic sector for mass separation. Operating conditions and analytical parameters were essentially those recommended by the manufacturer.

Ultrasonic Nebulizer. The ultrasonic nebulizer (USN) used in this study was provided by Applied Research Laboratories, En Vallaire, CH-1024, Ecublens, Switzerland. A peristaltic pump is used to deliver sample solution to the oscillating surface of the quartz plate. Aerosols having smaller and more uniform particle size are produced, and transported first to the heated tube, and then to the condenser by the nebulizer Ar gas. The water in the aerosol is mostly removed by this process, and only the dry aerosol is introduced into the plasma. The temperature of the heated tube and the condenser was kept at 120 C and 1 C respectively. The sample introduction rate was adjusted to 2 - 4 ml/min.

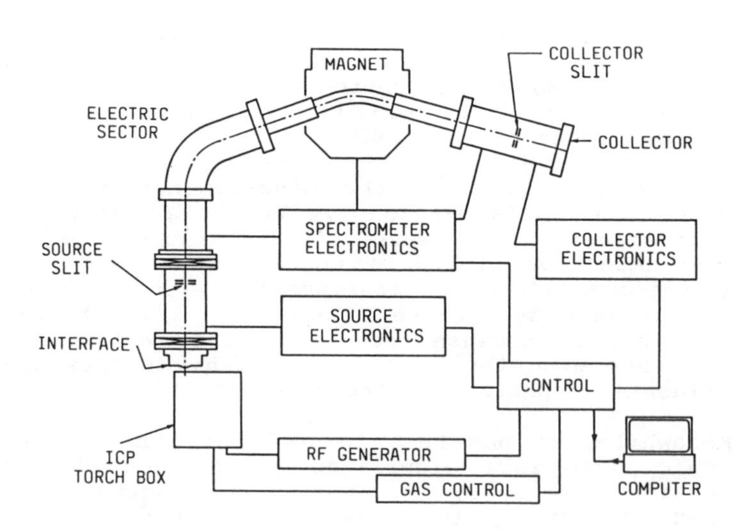

Figure 1 Schematic Diagram of High Resolution ICP-MS

3 RESULTS AND DISCUSSION

Ultrasonic Nebulizer

There have been several reports on the inadequacies of ultrasonic nebulization with desolvation. The main disadvantages were reported to be short as well as long-term instability, lack of reliability, and strong memory effects[7-8]. Accordingly, the above three points were fully examined to evaluate the suitability of this technique.

Long-term Stability. Long-term reproducibility obtained by 2 hours of measurements is shown in Table 2. Each measurement was carried out every 10 min, and a blank solution (1% HNO_3) was continuously introduced during the measurements. As the results in Table 2 show, the obtained relative standard deviations (RSD) were around 4% for all the elements examined.

Memory Effect. Figure 2 summarized residual memory effects. First a blank solution (1% HNO_3) was introduced, and then changed to a solution containing 1 ppm of Lu for about one minute and the intensity of the signals was measured. Immediately after the measurement, the standard solution was replaced with the blank solution, and the signals were monitored every 30 sec for 20 min. The signal intensity is expressed in terms of counts per second (CPS). The intensity of the standard solution at mass number 175 (^{175}Lu) was so strong that it was estimated from that of ^{176}Lu. As shown in Figure 2, CPS decreased rapidly and reached to less than 1/1000 of the standard solution within 30 sec. Accordingly, though the signal was still far more higher than that of the blank even after 20 min, it can be concluded from the above that there seem to be no serious memory effects if the CPS of the highest concentration sample is kept at less than 1000-10000 times that of the blank.

Table 2 Long-term Stability

Element	RSD (%)
Co	3.9
Y	3.4
La	3.2
Tb	3.5
Pb	3.7
Bi	3.8
U	4.1

Number of Measurements: 12
Measurements Interval: 10 min
Concentration of Elements: 10 ppb

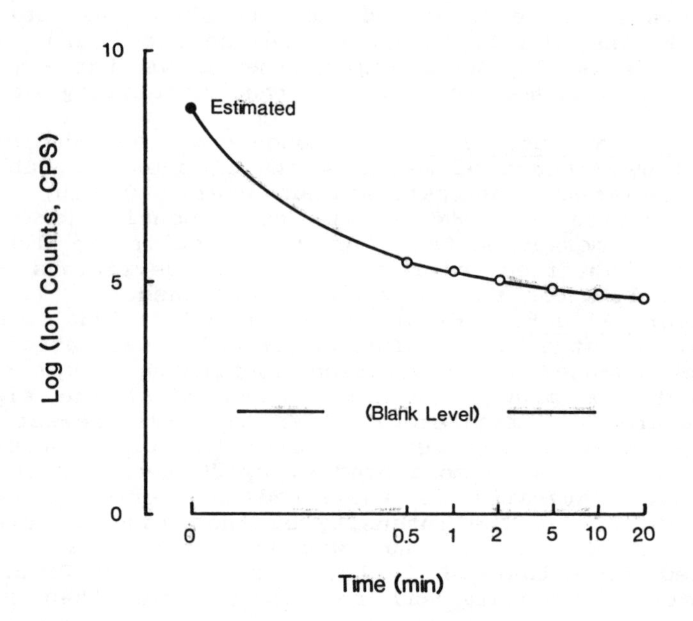

Figure 2 Residual Memory Effect

Table 3 Relative Sensitivity

Elements or Molecules	USN/Pneumatic	
	10 ppb	Blank
Mg	28	(17)
ArO	(1.6)	0.3
Co	20	1.7
Ar_2	0.97	0.92
In	14	1.3
La	10	3.1
Pb	20	8.1
Bi	18	5.1

Relative Sensitivity. The increase of the signal intensity due to the use of USN is summarized in Table 3. In a 10 ppb solution, the increase ranged from 10 to nearly 30 times for all the elements examined. A slight increase observed in ArO might be due to a small amount of Fe contained as impurities. Although the increase of signal intensity was also noticed in the blank, the magnitude of the increase was much smaller than that of the 10 ppb solution. The marked increase of Mg in the blank was proved to be due to impurities. The substantial decrease of ArO signal can be explained simply in terms of the removal of the main source of oxygen in plasma (water) by the desolvation process in USN.

Typical Detection Limit

The detection limits of several elements obtained by PlasmaTrace combined with USN are listed in Table 4, together with the integrated counts of the 10 ppt and the blank solution. Detection limits are defined here as the equivalent concentration of three times of the standard deviation of the blank response. The standard deviation of the blank solution was calculated from the 10 consecutive measurements carried out within relatively short periods (usually less than 5 min). The total peak dwell times were about 5 sec.

Figure 3 and 4 show the mass spectrum of 10 ppt of U and the blank solution as examples. While a beautiful symmetrical peak with very high signal intensity is observed even in a very low 10 ppt solution, the counts accumulated in a considerable number of channels are only one or two in the blank solution. These are the major reasons why it is possible to attain such extremely low detection limits as is shown in Table 4. All the other elements shown in Table 4 showed similar mass spectra.

Table 4 Typical Detection Limit

Element	Total Counts*		Detection Limit
	10 ppt	Blank	(ppq = pg/l)
Y	31,790	35.6	2.3
La	47,344	61.4	1.4
Nd	7,005	26.2	5.6
Tb	43,927	32.0	0.8
Th	32,072	32.6	1.0
U	44,326	29.6	0.7

* Average of 10 measurements.

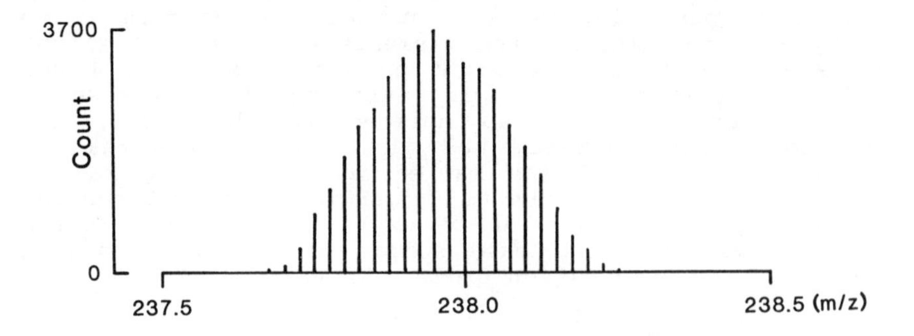

Figure 3 Mass Spectrum of 10 ppt of U

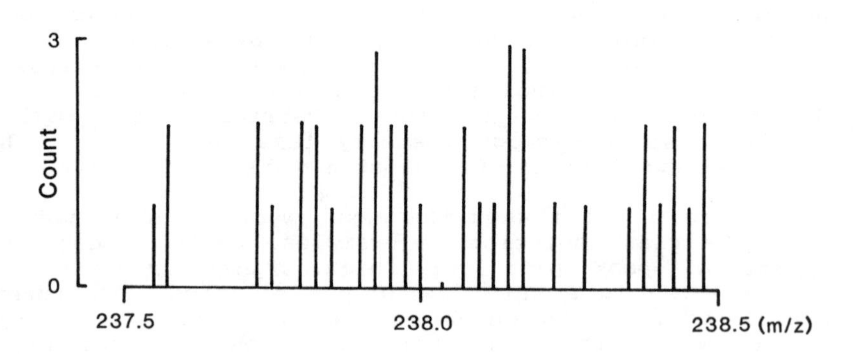

Figure 4 Mass Spectrum of the Blank Solution

Lineality of Calibration Curve

Figure 5 shows the calibration curves of Eu obtained by this extremely sensitive system as a typical example. Excellent linearity was observed in the range between 0 to 200 ppt. All the data points are practically superimposed on the regression line. This excellent linear range extended up to as high as nearly 200 ppb. However, the use of standard solutions of higher concentrations is considered to be not appropriate because, as mentioned above, the signal intensity of the highest standard solution should be less than 10,000 times of that of the blank solution to avoid errors due to the memory effects. Since good linearity was observed within a very wide concentration range, calibration curves are constructed usually by using only one standard solution (10 or 50 ppt) and the blank in most cases.

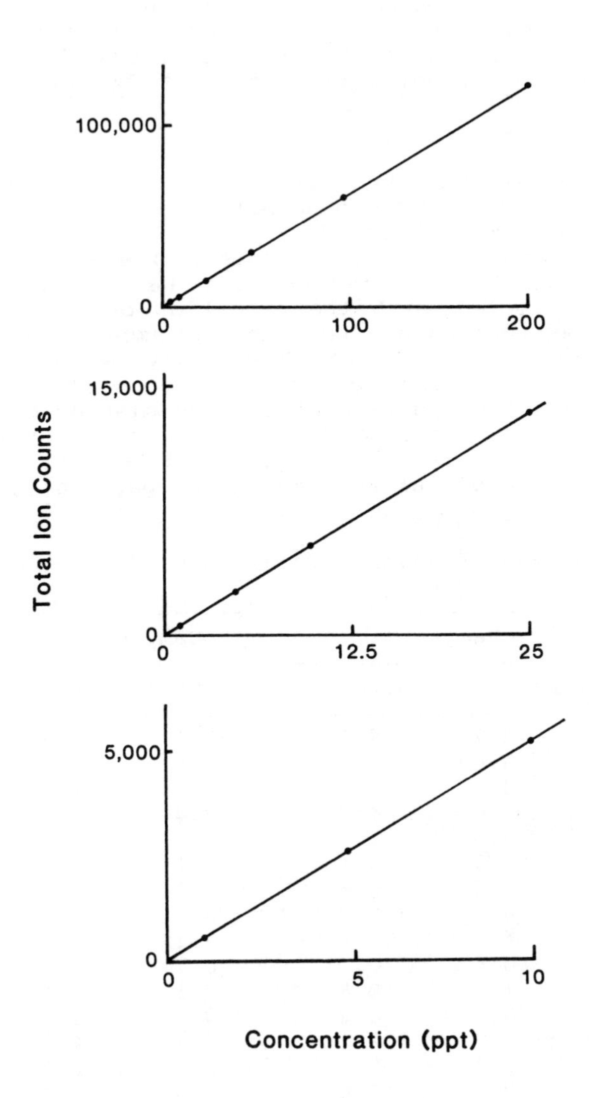

Figure 5 Typical Calibration Curve

Oxide Correction

The isobaric interference from BaO on Eu must be corrected mathematically because there is no available Eu isotopes which do not overlap with BaO. This oxide correction was done as follows. First, the contents of Sm were determined using both ^{147}Sm and ^{152}Sm, the latter overlaps with ^{136}BaO. Then the magnitude of the interference of ^{135}BaO on ^{151}Eu was estimated from the difference of the Sm content derived from ^{152}Sm and ^{147}Sm.

Comparison of Methods

Table 5 represents the results obtained by a quadrupole type ICP-MS after coprecipitation with $Fe(OH)_3$, and by the direct method developed in this study for two arbitrarily selected water samples. There are big differences between the two methods for La in sample 2 and Dy in sample 1. But with the exception of these two cases, it may be concluded that the agreements between both methods are acceptable for the analyses of these extremely low concentration levels. Although it is not possible to specify the causes of the above discrepancies, it seems to be most probable that the values obtained by the coprecipitation method are too high because these values also deviate from the well known REE pattern.

Table 5 Comparison of Results Obtained by Two Methods

Element	Sample 1		Sample 2	
	Fe ppt.[a]	Direct[b]	Fe ppt.[a]	Direct[b]
La	13.2	9.5	47.1	13.6
Ce	5.8	5.3	17.6	20.6
Pr	0.9	1.3	3.0	4.0
Nd	5.0	4.3	13.4	14.2
Sm	3.5	2.2	3.1	3.9
Eu	2.9	0.5	1.1	1.1
Gd	1.6	0.6	2.8	2.9
Tb	0.2	0.2	0.3	0.4
Dy	4.7	0.8	3.1	2.6
Ho	0.4	0.2	0.7	0.6
Er	1.5	0.8	2.1	2.4
Tm	0.3	0.2	0.4	0.5
Yb	2.0	2.1	3.2	4.2
Lu	0.4	0.5	0.7	1.1
Th	0.8	0.5	1.5	0.8
U	6.5	7.1	22.6	22.9

[a] Coprecipitation with $Fe(OH)_3$ and PlasmaQuad.
[b] Direct determination by PlasmaTrace plus USN.
Concentration unit is ng/l (ppt).

Mean and Range of REEs

Ranges and means of the lanthanide series elements in river water, randomly sampled at various places in Japan, are presented in Table 6 together with those of Th and U. The concentrations of these elements range so widely that the ratios of the highest values to the lowest values are more than 100 for most of the elements listed in Table 6. In spite of this widely distributed concentration range, the mean values of each lanthanide element in the water sample also clearly show similar characteristic tendencies of being higher in the element with the even atomic number than the next element with the odd atomic number, often observed in rock and soil samples.

Histogram of REEs

It was found that the frequency distributions of the concentrations of all the lanthanide series elements in river water were highly asymmetric (positively skewed). When the values of REEs were log-transformed, however, it was possible to obtain the histograms which appeared to be very close to those of normal distributions. The results of La, Nd and Yb are shown in Figure 6 as typical examples. Thorium and U also showed similar tendency, though their histograms are not shown here.

Table 6 Rare Earth Elements in River Water (ppt)

Element	No. of Samples	Range	Mean
La	77	0.53 - 214	22.2
Ce	77	1.9 - 442	35.2
Pr	77	0.0[4] - 65	6.2
Nd	77	2.5 - 335	28.1
Sm	77	0.0[4] - 66.2	5.3
Eu	77	0.0[4] - 15.8	1.8
Gd	77	0.0[4] - 95.4	8.1
Tb	77	0.0[4] - 11.8	1.0
Dy	77	0.7 - 63.1	6.1
Ho	77	0.2 - 10.6	1.4
Er	77	0.6 - 27.2	4.8
Tm	77	0.0[4] - 5.1	0.89
Yb	77	0.34 - 37.4	6.0
Lu	77	0.15 - 8.4	1.3
Th	73	0.0[4] - 32.5	3.5
U	77	0.87 - 199	25.1

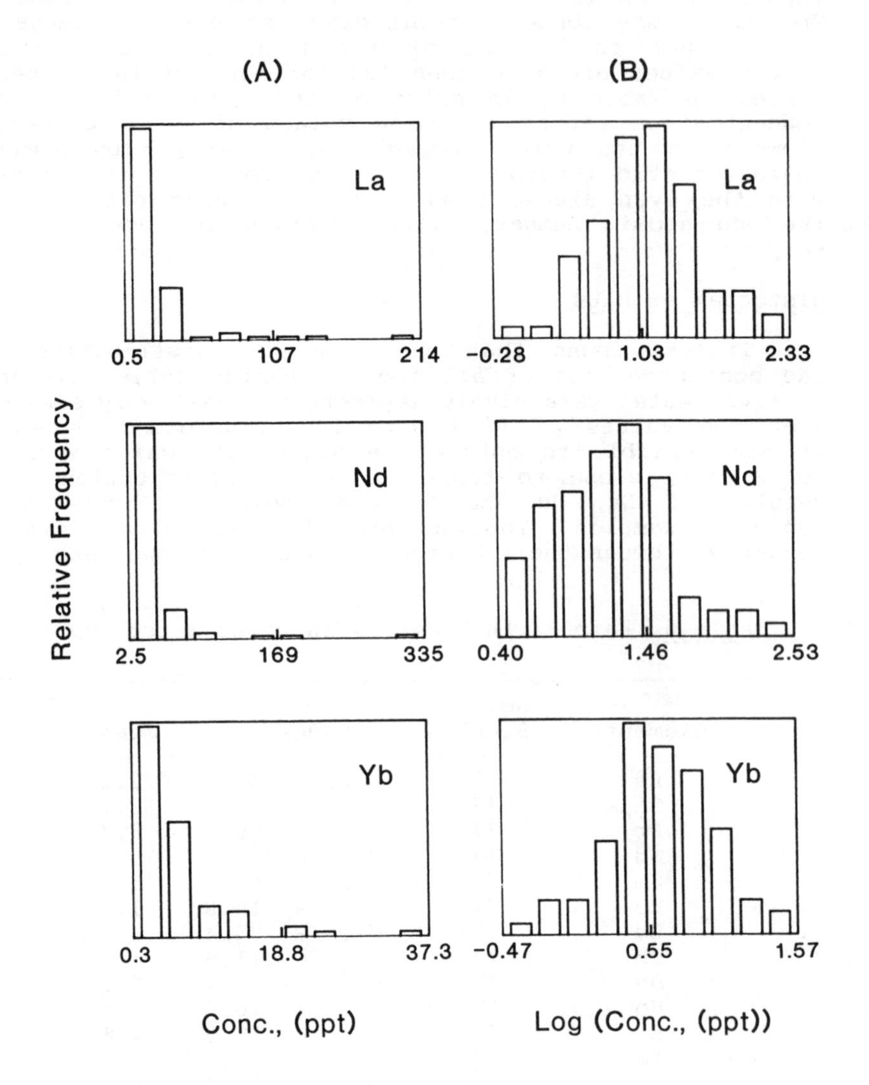

Figure 6 Histograms Showing the Distribution of Some REEs Before (A) and After (B) Log-transformation

4 SUMMARY

Ultra-trace levels of REEs in terrestrial water were determined by high resolution ICP-MS with an ultrasonic nebulizer. The proposed method was proved to be extremely sensitive and, therefore, it was possible to determine most of the REEs in water samples directly with acceptable speed and accuracy. Results obtained by this technique compared favorably with those by the more complicated coprecipitation procedure with $Fe(OH)_3$.

REFERENCE

1. A. Tsumura and S. Yamasaki, in preparation.
2. N. Bradshaw, E. F. H. Hall and N. E. Sanderson, J. Anal. At. Spectrom., 1989, 4, 801.
3. M. Morita, H. Ito, T. Uehiro and K. Otsuka, Anal. Sci., 1989, 5, 609.
4. K. W. Olsen, W. J. Haas, Jr., and V. A. Fassel, Anal. Chem., 1977, 49, 832.
5. V. A. Fassel and B. R. Bear, Spectrochim. Acta, 1986, 41B, 1089
6. R. J. Thomas and C. Anderau, Atom. Spectrosc, 1989, 10, 71.
7. R. P. Browner and A. W. Broon, Anal. Chem., 1984, 56, 875A.
8. M. Thompson and J. N. Walsh, 'Handbook of Inductively Coupled Plasma Spectrometry', Blackie, Glasgow, 1989.

The Application of Isotope Dilution Techniques to the Accurate and Precise Determination of Lead in Reference Materials by ICP-MS

M.J. Campbell, C. Vandecasteele, and R. Dams

LABORATORY OF ANALYTICAL CHEMISTRY, UNIVERSITY OF GHENT, INSTITUTE FOR NUCLEAR SCIENCES, PROEFTUINSTRAAT 86, B-9000 GHENT, BELGIUM

1 INTRODUCTION

Isotope dilution analysis coupled with thermal ionisation mass spectrometry is a powerful technique capable of highly accurate and precise determinations of elements that have two or more stable isotopes.[1] The principle of the method is that by altering the natural ratio between two isotopes in the sample, by the addition of an accurately known quantity of an isotopic spike, and measuring the ratio of the mixture, the concentration of analyte present in the original sample can be deduced from a knowledge of the weight of sample taken, the quantity of spike added and the ratio of the isotopes in the original sample, spike and mixture.

The two major advantages of the method over conventional analysis against a set of standards stem from the fact that ratios, rather than absolute sensitivities, are measured. Firstly, once the spike has been added to the sample and allowed to come to equilibrium , losses of analyte become unimportant since the isotopic ratio of the mixture would not be altered. The second advantage is that the determination is largely unaffected by changes in instrumental sensitivity and matrix effects since both isotopes should be affected to the same extent and so the ratio will remain constant. The use of isotope dilution with thermal ionisation mass spectrometry is capable of providing very accurate and highly precise analyses with typical RSDs of less than 0.1%. However the technique has some drawbacks. Usually the analyte has to be separated from its matrix and then mounted as uniformly as possible on the filament. Clearly this procedure often involves considerable sample pretreatment and chemical separations, is time consuming and increases the risk of contamination. The analysis time is relatively long, often taking several hours and consequently sample throughput rates are low , even if multiple filament devices are used.

The quadrupole mass analyser used in ICP-MS separates ions on the basis of their mass to charge ratio thereby providing isotopic composition data for all elements present in the sample. The method of isotope dilution analysis is therefore directly applicable to ICP-MS, thereby providing us with another hyphenated variant: ID-ICP-MS. The advantages of ID-ICP-MS over thermal ionisation IDMS are firstly that sample pretreatment, other than dissolution, is not usually required and secondly the sample handling times are of the order of ten minutes for introduction of a sample, five replicate measurements, washout and introduction of the subsequent sample which allows a much higher sample throughput than is possible with TIMS. We have developed suitable methodologies to determine lead concentrations in reference materials using ID-ICP-MS which are presented later.

Lead has four stable isotopes ^{204}Pb, ^{206}Pb, ^{207}Pb and ^{208}Pb of which three are the daughter nuclides of radioactive decay from uranium and thorium. Consequently the isotopic composition of a given ore sample will be dependent on its age and the relative proportions of thorium and uranium in the parent strata. Typically, the abundance of ^{208}Pb in an ore sample is about 50% whereas the ^{206}Pb is nearer 25% so the $^{208}Pb:^{206}Pb$ ratio is about 2. The isotope dilution approach for the analysis of lead is illustrated on figure 1. Figure 1a shows the isotopic composition for lead in a plankton sample in which the $^{208}Pb:^{206}Pb$ ratio is 2.085. The peaks at 203 and 205 mass units are due to thallium which was incorporated as an internal standard in all solutions to enable accurate blank subtractions to be made. Figure 1b shows the spectrum for the synthetic spike NBS 983 which has been enriched with the minor lead isotope ^{206}Pb such that its abundance is 92% and the abundance of ^{208}Pb is 1%. The spectrum for lead in the plankton sample after the addition of the spike is shown in figure 1c. It is imperative to add sufficient spike to alter the natural ratio of the analyte significantly. The quantity of spike added was chosen to give a $^{208}Pb:^{206}Pb$ ratio of approximately unity in order to get the best counting statistics on the ratio.

A simple treatment of the underlying principle of the isotope dilution technique and derivation of the equation used to calculate the concentration of analyte in the original sample was presented by Longerich.[2]

In ICP-MS analyses ions are not transmitted through the quadrupole mass filter with equal efficiencies; therefore the observed isotopic ratios must be corrected for the instrumental fractionation to give the "true" ratio. Consider an element, E, with major isotope ^aE and minor isotope ^bE having a (natural) ratio R_s. Let the number of moles of each isotope be denoted by ^aX and ^bX respectively. The measured ratio, R_s, is equal to the ratio of the observed count rates due to detection of ions from ^aX and ^bX in the sample, multiplied by a sensitivity factor for each nuclide such that:

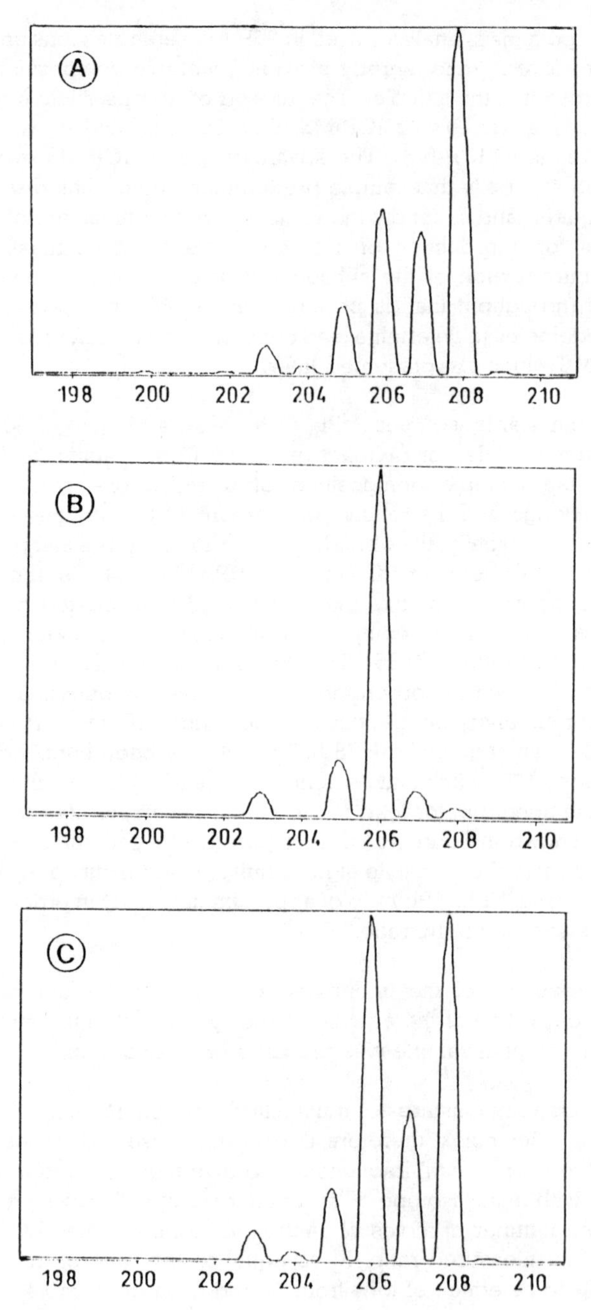

Figure 1 Spectra for (A) (natural) lead in Plankton, (B) NBS 983 lead spike, (C) mixture of lead from spike and sample.

$$R_S = \frac{{}^aS\,{}^aX_s}{{}^bS\,{}^bX_s} \tag{1}$$

Where ${}^aS$ and ${}^bS$ are the respective sensitivity factors for nuclides a and b (in units of counts.s^{-1}.mol^{-1}). Rearranging (1):

$${}^aS\,{}^aX_s = R_s\,{}^bS\,{}^bX_s \tag{2}$$

A similar expression can be written for R_t, the ratio in the spike solution, enriched to a high purity with the minor abundant isotope, ${}^bX_t$.

$${}^aS\,{}^aX_t = R_t\,{}^bS\,{}^bX_t \tag{3}$$

When the spike isotope is added to the sample, the ratio in the mixture, R_m, is given by:

$$R_m = {}^aS\,{}^aX_s + {}^aS\,{}^aX_t\,R\,{}^bS\,{}^bX_s + {}^bS\,{}^bX_t \tag{4}$$

Substituting equations (2) and (3) into equation (4) the sensitivity factor for isotope a is eliminated.

$$R_m = \frac{R_s\,{}^bS\,{}^bX_s + R_t\,{}^bS\,{}^bX_t}{{}^bS\,{}^bX_s + {}^bS\,{}^bX_t} \tag{5}$$

Rearranging equation (5) and multiplying it out, the sensitivity factor for isotope b can be cancelled out:

$$R_m\,{}^bX_s + R_m\,{}^bX_t = R_s\,{}^bX_s + R_t\,{}^bX_t \tag{6}$$

Rearranging equation (6) and solving it for ${}^bX_s$ (the moles of isotope b in the sample):

$$^{b}X_s = \frac{^{b}X_t(R_m - R_t)}{(R_s - R_m)} \qquad (7)$$

It is clear from the above derivation that the relative fractionation of the two isotopes due to instrumental bias does not enter into the calculation for the number of moles of the minor isotope present in the sample. However, if the total number of moles of analyte in the sample is to be determined, a knowledge is required of the relative atomic mass of the analyte and the fractional composition of the minor isotope. If these are not known (owing to variations in nature, as is the case for lead) they must be determined from the observed ratio data.

The concentration of analyte present in the natural sample can be calculated from the following general formula:

$$M_s = \frac{^{b}X_t A(R_m - R_t)}{^{2}f(R_s - R_m)} \qquad (8)$$

Where M_s is the mass of analyte present in the sample
$^{b}X_t$ is the number of moles of spike isotope 2 (tracer) added
A is the relative atomic mass of the element (in the sample)
R_m is the isotopic ratio of the mixture
R_t is the isotopic ratio of the spike (tracer)
R_s is the isotopic ratio of the sample
^{2}f is the isotopic abundance of isotope 2 (used for spiking) in the natural sample.

Usually, the concentration of the analyte is expressed per gramme of sample by dividing M_s by the mass of sample taken, corrected for its water content. A statistical assessment of the errors associated with isotope dilution analysis was presented by De Bievre and Debus.[3]

Since the technique was applied to the analysis of reference materials with well characterized lead concentrations, the spike addition was determined from a knowledge of the quantity of lead present in the sample taken and an assumption of the likely lead isotopic composition in the sample (based on the NBS 981 "natural" lead standard). It was established from the literature[4] that optimum performance of the ID technique could be achieved over a range of spiked ratios from 0.1 to 10 and the approach used gave ratios between 0.9 and 1.1 for all samples analysed. (NB the true isotopic composition and abundances were determined from the data obtained during analysis for use in the final calculations).

2 EXPERIMENTAL

Initial experiments were conducted to assess the suitability of a closed vessel, microwave digestion procedure using nitric acid.[5,6] This approach offered the benefits of rapid sample preparation (less than 20 minutes) and reduced contamination; however for some of the reference materials studied a precipitate was observed, probably due to the presence of silicates in the material. Using this technique it was probable that the organic matrix was only partially destroyed and the use of perchloric acid, to completely destroy the matrix, was precluded because of the risk of an explosion. The use of hydrofluoric acid to remove silicon from the matrix as the volatile silicon tetrafluoride would require evaporation and so the primary advantage of closed vessel digestion would be lost. Since it was likely that analyte could be trapped within the interstices of the precipitate or could be co-precipitated onto its surface, this technique was abandoned for the time being.

The sample dissolution procedure finally adopted for all materials studied involved an open mixed acid digestion in Teflon beakers. A given sample was analysed in two separate batches. Each batch consisted of four portions of sample plus a blank, three of the portions were appropriately spiked and the fourth portion was analysed without any spike to obtain the natural lead isotope ratio of the sample. The samples (0.1g) were weighed by difference in plastic weighing boats (≈0.4g) and transferred into the acid washed Teflon beakers. Nitric acid was added (3ml, sub-boiling distillation of high purity material) and the beakers were transferred to a hot plate and allowed to digest for one hour to destroy most of the organic matrix. The samples were allowed to cool and hydrofluoric and perchloric acids were added (3ml of each) and the samples were returned to the hot plate and allowed to digest overnight. The samples were heated to near dryness to drive off excess hydrofluoric acid and a further aliquot of nitric acid was added. The spike was added after sample dissolution was complete to ensure sample and spike were equilibrated. The solutions were transferred to appropriate acid washed volumetric flasks and diluted with deionised water to give a final (spiked) total lead concentration of approximately $100 \mu g l^{-1}$. The final solution also contained thallium ($10 \mu g l^{-1}$) as an internal standard and a surfactant, Triton X-100 (0.1%v/v), was added to improve solution transport and nebulisation efficiencies and reduce memory effects.

A portion (1g) of each reference material analysed was dried in an oven at 110°C to constant mass with the appropriate batch analysis, to characterize its water content.

Experiments were conducted to evaluate the dwell and total acquisition times for optimum isotope ratio precision and it was concluded that a $320 \mu s$ dwell time with a scan duration of 90 seconds gave the best results. The instrumental parameters are presented in table 1.

Table 1 Scanning and instrumental parameters

Dwell Time	320 μs
Number of Scans	550
Memory Allocation	512 Channels
Mass Range	197-210 amu
Total Acquisition Time	90 s
Coolant Gas Flow	~12 $lmin^{-1}$
Auxiliary Gas Flow	~0.5 $lmin^{-1}$
Nebuliser Gas flow	~0.75 $lmin^{-1}$
Forward Power	1.35 kW
Reflected Power	<5 W

In order to ensure the quality of data produced in an analytical run, samples (in groups of 4 or 6) were analysed between two pairs of isotope ratio standards (NBS SRM 981 and 983) and only if the ratios determined for the standards were within ±0.5% of their designated values was the data accepted. The following analytical protocol was adopted for all analyses:

(i) Reagent blank
(ii) NBS 981 (50μgl^{-1}) Calibrant
(iii) NBS 981 Test
(iv) NBS 983 (50μgl^{-1})
(v) Procedural blank A
(vi) Procedural blank B
(vii) Sample A 1-4
(viii) NBS 981
(ix) NBS 983
(x) Sample B 1-4
(xi) NBS 981
(xii) NBS 983

Five measurements were made on each solution and then a wash solution was run. Care was taken to ensure that the analyte signal had returned to background level before a subsequent analysis was begun. On a second occasion, samples prepared in the second batch would be analysed using the same sequence.

In order to produce a final result for a material, a total of six spiked and two unspiked samples were analysed together with appropriate blanks and standards as outlined above.

3 RESULTS AND DISCUSSION

The results accepted and used by the EEC Community Bureau of References (BCR) to certify the lead content of a River sediment material, BCR 320 are presented on figure 2. A total of 15 sets of results were reported[7], but only the seven values to the left of the figure were considered to be sufficiently accurate and precise to be used for certification. The certified value, on the extreme right of figure 2, for lead in BCR 320 is 42.3±1.6 $\mu g g^{-1}$ and the value determined by ID-ICP-MS was 41.7±0.79 $\mu g g^{-1}$. The precisions of the three isotope dilution measurements are superior to those of the other techniques and the ID-ICP-MS precision lies between the two TI-IDMS values which were obtained by two separate laboratories. Table 2 gives a summary of results for the analysis of lead in various reference materials using ID-ICP-MS and compares the values against the certified and TI-IDMS results. The precisions obtained by ICP-MS and TIMS are comparable. The agreement observed between the ID-ICP-MS result and the certified value is good for all reference materials analysed, with the exception of CRM 277 and 280 (river and lake sediments respectively) for which our results were significantly lower than the certified values. It should be stressed that no residue was visible to the naked eye after the original sample preparation procedure and that analysis of three separate batches of the materials gave reproducible results. In light of these discrepancies, the analysis of BCR 277 was repeated using an alternative high pressure (mixed acid) dissolution technique in a closed Teflon vessel placed in a Parr bomb. This approach ought to ensure total dissolution of the sample and minimise risk of analyte loss. The result obtained was 143.6$\mu g g^{-1}$ Pb which falls within the certified 95% confidence interval for this material. This example serves to highlight a potential pitfall of the technique: it is imperative that all the analyte present in the sample be brought into solution and allowed to equilibrate with the spike. If this consideration is met, the isotope dilution

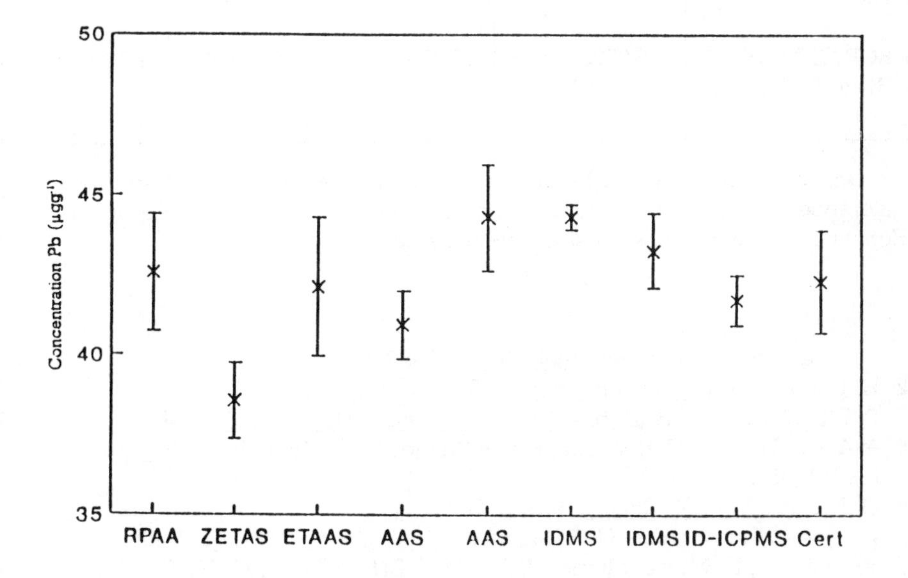

Figure 2 Certification results for lead in BCR CRM 320 (River sediment). Error bars show 95% confidence limits.

technique is a very powerful tool for trace element analysis and is clearly compatible with ICP-MS, affording a better sample throughput rate and considerably less sample pretreatment than isotope dilution analysis using thermal ionisation mass spectrometry.

Table 2 A comparison of our ID-ICP-MS results for lead in certain BCR reference materials with TI-IDMS results for the certification exercises and the final certified results.

Sample	ICP-MS				
	Natural ratio $^{208}Pb:^{206}Pb$	n	ID-ICP-MS	TI-IDMS	Certified Value
BCR CRM 60 (Aquatic Plant)[8]	2.0864	6	64.08±1.44	64.80±0.37	63.80±3.20
BCR CRM 61 (Aquatic Plant)[8]	2.1089	6	62.36±0.67	61.11±1.45	64.40±3.50
BCR CRM 62 (Olive Leaves)[8]	2.1390	6	25.83±0.43	26.41±0.46	25.00±1.50
BCR CRM 320 (River Sediment)[7]	2.0923	6	41.70±0.79	44.32±0.39	42.30±1.60
BCR CRM 277 (Estuarine Sediment)[7]	2.0968	9	137.70±1.4	148.10±0.90 145.00±3.04 134.70±0.57	146.00±3.00
BCR CRM 280 (Lake Sediment)[7]	2.0923	9	76.35±1.10	82.70±0.41 81.30±0.92 72.30±1.41	80.20±2.30

One of us, MJC, would like to acknowledge the Community Bureau of Reference for providing financial support for this work through the European Economic Community Sectoral Grants Scheme.

REFERENCES

1. J. D. Fassett, P. J. Paulsen, Anal. Chem., 1989, 59, 643a.
2 H. P. Longerich, Atom. Spectrosc., 1989, 10, 112.
3 P. J. De Bièvre, G. H. Debus, Nucl. Instr. and Meth., 1965, 32, 224.
4 A. A. van Heuzen, T. Hoekstra, B. van Wingerden, J. Anal. Atom. Spectrosc., 1989, 4, 483.
5 G. Knapp, Tr. Anal. Chem., 1984, 3, 182.
6 P. Aysola, P. Anderson, C. H. Langford, Anal. Chem., 1987, 59, 1582.
7 B. Griepink, H. Muntau, Report EUR 11850 EN, ECSC-EEC-EAEC, Luxemburg (1988).
8 E. Colinet B. Griepink, H. Muntau, Report EUR 8119 EN, ECSC-EEG-EAEC, Luxemburg (1982).

Effect of Digestion Methods on ICP-MS Measurements of Be, V, Ga, Sb, La, Bi in Organic Fertilizers

A. Kawasaki, K. Haga, S. Arai, and S. Yamasaki

NATIONAL INSTITUTE OF AGRO-ENVIRONMENTAL SCIENCES, 3-1-1 KANNONDAI, TSUKUBA, IBARAKI, JAPAN 305

1 INTRODUCTION

There are many kinds of rare metals that are increasingly used in high-technology industries. The possible effects of these metals on the ecosystem are worrisome because most of these metals are unfamiliar to our living environments and their biological effects are not sufficiently known. Agricultural materials made from wastes can be a source of contamination in the farmlands. The amount and distribution of rare metals in them, however, are scarcely elucidated.

ICP-MS is considered to be suitable for analysis of agricultural materials because of its multielemental, sensitive and wide dynamic range characteristics.

The objectives of this study are to examine and select a digestion method for the determination of rare metals in sewage sludge and compost using ICP-MS. Six elements, beryllium, vanadium, gallium, antimony, lanthanum and bismuth, were investigated regarding their contents and recovery. The amounts of four other elements, nickel, copper, cadmium and lead were also determined and compared with the certified values.

2 MATERIALS AND METHODS

Materials

Sewage sludge with organic coagulant (No.101), sewage sludge with inorganic coagulant (No.102) and a compost of pig waste with sawdust (No.103) were used. These materials were the reference materials that had been prepared for the determination of common heavy metals (Table 1).

Digestion Methods

Dry ashing (Method A). An approximately 5 g sample was weighed accurately in a Pt dish and ignited at low

Table 1 Certified Values of Reference Materials

	No.101 (Sludge)		No.102 (Sludge)		No.103 (Compost)	
	mg/kgDM	(Range)	mg/kgDM	(Range)	mg/kgDM	(Range)
Ni	26.0	2.0	38.8	4.6	43.1	2.4
Cu	170	4	352	17	143	3
Cd	2.68	0.16	4.75	0.35	2.31	0.13
Pb	58.4	7.6	207	29	28.1	26.2

red heat to carbonize it. The contents were transferred to a 300 ml tall beaker with H_2O and 10 ml HCl was added and heated. After cooling, the solution was diluted exactly to 250 ml (1).

HNO_3-$HClO_4$ digestion (Method C). An approximately 5 g sample was weighed accurately in a 300 ml tall beaker and 30 ml HNO_3 was added then boiled. Most of the acids were evaporated to dryness and allowed to cool. 30 ml HNO_3 and 10 ml $HClO_4$ were added and digested. Excess acids were evaporated to dryness and were allowed to cool. 25 ml HCl (1+5) was added and heated. After cooling, it was diluted exactly to 100 ml (2).

HNO_3-HCl digestion (Method D). An approximately 5 g sample was weighed accurately in a 300 ml tall beaker and 5 ml HNO_3 and 15 ml HCl were added and heated. Addition of the acids and heating were repeated until the sample was digested completely. 25 ml HCl (1+5) was added and heated. After it cooled, it was diluted exactly to 100 ml (3).

6M HCl under reflux (Method E). An approximately 5 g sample was weighed accurately in a 250 ml Erlenmeyer flask and 20 ml 6M HCl was added and heated with a reflux condenser for 1 hour. After the condenser was washed with 0.1M HCl, the contents were filtered and diluted exactly to 100 ml (4).

HNO_3 decomposition in a pressurized vessel (Method F). An approximately 0.5 g sample was weighed accurately in a Teflon beaker and 5 ml HNO_3 was added and mixed. The Teflon beaker was set into a stainless steel vessel and heated at 120°C for 1 hour. After cooling, the solution was diluted exactly to 50 ml (5).

HNO_3-$HClO_4$-HF digestion (Method H). An approximately 1 g sample was weighed accurately in a 100 ml Teflon beaker and 25 ml HNO_3 and 10 ml $HClO_4$ were added and heated. After most of nitric acid was evaporated, the same quantity of HNO_3 was added and heated to digest them again. After it cooled, 5 ml $HClO_4$ and 10 ml HF were added and heated. After most of the acids were evaporated, 10 ml HF was added and heated again. After most of the acids were evaporated, 5 ml HNO_3 was added and heated. After it cooled, it was diluted exactly to 100 ml (6).

Five replicates of each sample were performed by each method.

Recovery Test

A recovery test was done by the adding of the standard solution to the samples in each method. These quantities were 2 µg Be, 30 µg V, 10 µg Ga, 10 µg Sb, 30 µg La, 10 ug Bi per 1 g dry matter respectively. This standard solution was prepared from the standard solution for atomic absorption spectroscopy (Wako Pure Chemical Industries, Ltd.).

In order to examine the effect of the inorganic and organic matter on recovery of Sb, 200 mg ferric hydroxide, 200 mg alumina (PF-254, Merck), 200 mg silica-gel (Wako-gel C-200, Wako Pure Chemical Industries, Ltd.) and 200 mg lignin (dealkaline, Tokyo Kasei Kogyo Co., Ltd.) were added to the 10 ug Sb standard solution separately or collectively, and then the recovery of Sb was measured by the HNO_3-$HClO_4$ digestion method. The precipitation which remained after HNO_3-$HClO_4$ digestion was digested by the HNO_3-$HClO_4$-HF method and Sb was measured in the precipitation. In order to examine the effect of the amount of silicate on recovery of Sb, 0, 50, 100, 200 or 500 mg silica-gel was added to 10 µg Sb, and the recovery of Sb from the digested solution and the precipitation was also measured.

ICP-MS Measurement

The sample solution was diluted to an adequate concentration with 1% HNO_3 and nebulized into ICP-MS (VG PlasmaQuad PQ-1, VG Elemental Ltd., Winsford, Cheshire, England), and determined simultaneously. The instrumental conditions for the ICP-MS measurements are listed in Table 2.

Table 2 Instrumental conditions for ICP-MS

Selected isotopes:	Be(9),	V (51),	Ni(60)
	Cu(63),	Ga(71),	Cd(111)
	In(115),	Sb(121),	La(139)
	Pb(208),	Bi(209)	
Internal standard:	100 µg In/l		
Ion source (ICP) conditions:			
Forward power	1300 W		
Reflected power	< 10 W		
Coolant flow	14.0 l/min		
Auxiliary flow	0.5 l/min		
Nebulizer flow	0.8 l/min		
Mass spectrometer:	Quadrupole mass filter		
Sampler:	1.0 mm orifice		
Skimmer:	0.75 mm orifice		

3 RESULTS AND DISCUSSION

Rare Metal Contents and Recovery

Rare metal contents and recovery measured by each digestion method and their standard deviation are shown in Figure 1.

Beryllium. The observed value ranges of Be measured by methods A, E and F were 0.09-0.12 ppm (No.101), 0.25-0.27 ppm (No.102) and 0.12 ppm (No.103). The value ranges measured by methods C, D and H, however, were 0.13-0.14 ppm (No.101), 0.31-0.34 ppm (No.102) and 0.13-0.19 ppm (No.103). The latter values were higher than those in methods A, E and F. So methods A, E and F were considered to be insufficient digestion methods for Be. Recoveries ranged from 80 to 120 % in every method except methods C and D of No 103. These recoveries were considered to be more than 120 % because the drift of the equipment was too much to compensate for.

Vanadium. In methods C and D, the observed values of V were 230 ± 27 ppm (No.101, method C), 96 ± 5 ppm (No.103, method C), 51 ± 11 ppm (No.101, method D) and 58 ± 12 ppm (No.103, method D). These values were extremely high. The sample solutions in methods C and D contained chlorine. So the mass peaks of the V and ClO were considered to overlap each other. The solutions digested by methods A and E contained Cl also, so the observed values might have been affected by ClO. Recoveries calculated only for methods F and method H, which were free from Cl, ranged from 80 to 120 %.

Gallium. The observed value ranges of Ga measured by methods A, E and F were 0.3-0.8 ppm (No.101), 1.5-2.2 ppm (No.102) and 0.7-2.2 ppm (No.103). These values were lower than measured by methods C, D and H, which ranged from 2.0 to 2.3 ppm (No.101), 3.2-5.2 ppm (No.102) and 2.7-4.0 ppm (No.103). So the extraction efficiencies of methods A, E and F were considered to be inferior. Recoveries ranged from 80 to 120 % except in methods A, E and F.

Antimony. Recoveries of Sb digested by method H were about 100 %. Other methods, however, showed very low recoveries, which were less than 10 % by method A, 10-40 % by methods C,E and F, and 50-80 % by method D. It was thought that organic or inorganic components in the samples affected the recovery of Sb because the recovery of blanks, or standard solution without samples, were about 100 % in all methods. The effects of components coexisting and the recovering method were investigated and will be described below.

Lanthanum. The observed values of La prepared by method H were highest among all methods. They were 3.9 ppm (No.101), 6.9 ppm (No.102) and 2.1 ppm (No.103). The digestive ability of method H for La was considered to be superior to the other methods. Recoveries of La ranged from 80 to 120 % in all methods.

Bismuth. The observed value ranges of Bi were 3.4-4.4 ppm (No.101), 5.9-7.9 ppm (No.102) and 0.03-0.06 ppm (No.103) in all methods except method A. Recoveries of

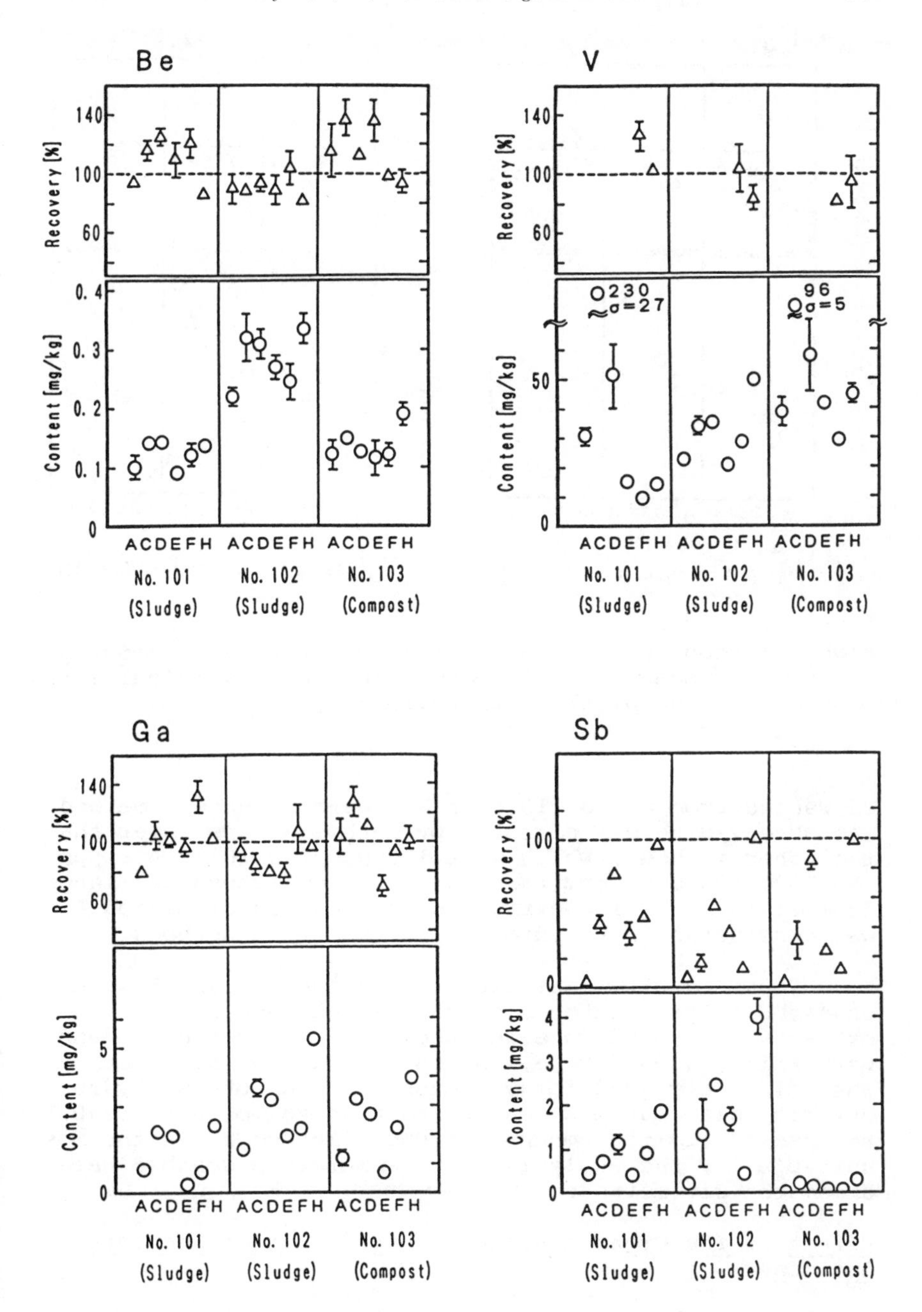

Figure 1 Rare metal contents and recovery.
○ Mean content; △ mean recovery; vertical lines, range of standard deviations.

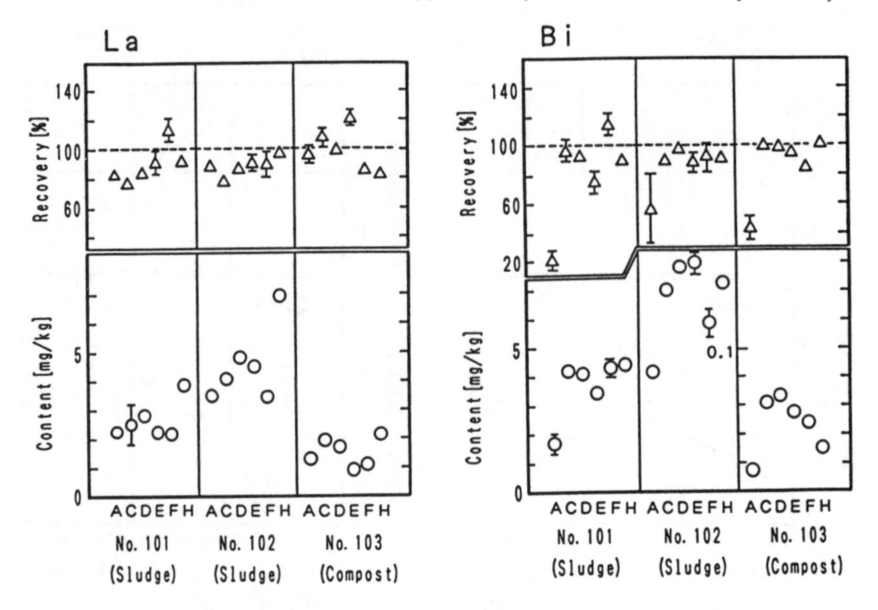

Figure 1 (continued) Rare metal contents and recovery. ○ Mean content; △ mean recovery; vertical lines, range of standard deviations.

Bi ranged from 80 to 110 % in all methods except method A. The observed values digested by method A were lower than the other methods, which were 1.6 ppm (No.101), 4.1 ppm (No.102) and 0.02 ppm (No.103), and low recoveries ranged from 21 to 56 %. The extraction efficiency of method A was considered to be inferior to the other methods.

These results indicated that methods A and E were unsuitable for rare metals analysis because of low extractability and interference with ClO. Methods C and D were fairly good for Be, Ga, La and Bi analysis. Method F was relatively good for V because it did not use hydrochloric acid, but was considered to give low extractability because of limited amounts of HNO_3. Method H was the most suitable for the analysis of rare metals described here among the six methods.

Effect of the Organic or Inorganic Matter on Recovery of Antimony

It was shown that the coexistent materials in the Sb solution reduced the recovery of Sb. Then, the effects of ferric hydroxide, alumina, silica-gel and lignin on the recovery of Sb were investigated. Bajo et al.(7) showed that Sb was adsorbed on the wall of the glass vessels and that the formation of insoluble Sb compounds were responsi-

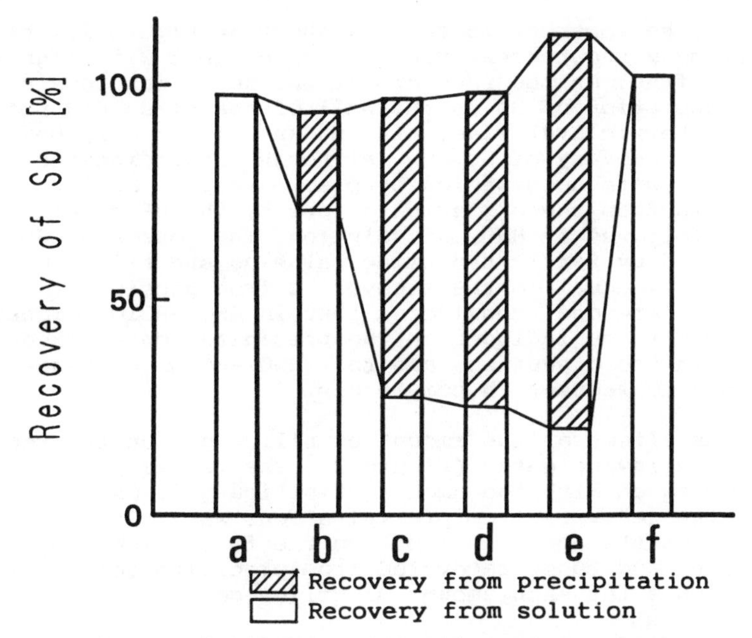

Figure 2 The effect of $Fe(OH)_3$, alumina, silica-gel and lignin on the recovery of Sb. (a) 10μg Sb solution; (b)+200mg $Fe(OH)_3$; (c)+200mg alumina; (d)+200mg silica-gel; (e)+200mg $Fe(OH)_3$, +200mg alumina and +200mg silica-gel; (f)+200mg lignin.

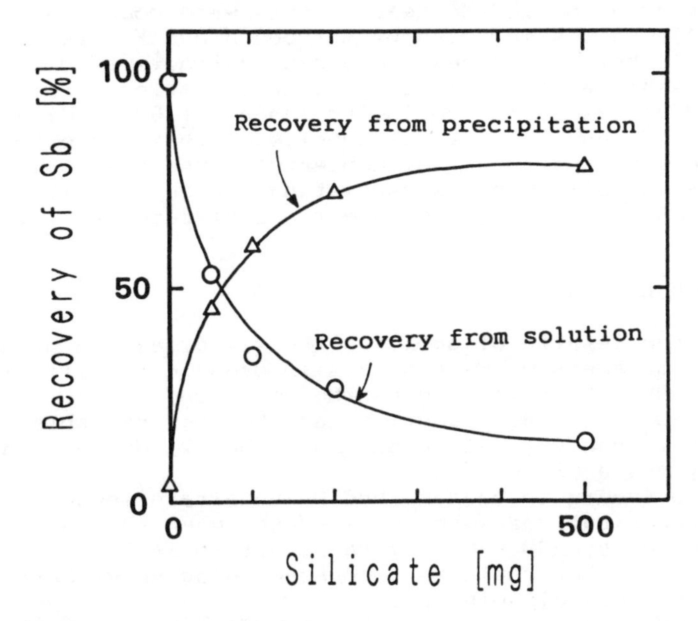

Figure 3 The effect of the amount of silica-gel on the recovery of Sb. ○ Recovery from solution; △ recovery from precipitation.

ble for the apparent losses. As shown in Figure 2, the treatment without coexistent matter or that with lignin had no effect on the recovery of Sb, so it was considered that adsorption of Sb on the wall of the glass did not happen. Ferric hydroxide, alumina and silica-gel, however, made the recovery of Sb low. Although these digested solutions were accompanied by precipitation, Sb was recovered from these precipitations by the HF digestion method followed by HNO_3 dissolution. The losses of Sb influenced by ferric hydroxide, alumina and silica-gel were nearly equal to the recoveries from precipitations. It was, therefore, considered that in HNO_3-$HClO_4$ digestion, Sb tended to be included in the precipitation of oxide formed in the digestion, and that HNO_3-$HClO_4$-HF digestion was able to recover Sb completely.

The effect of the amount of silica-gel on the recovery of Sb was investigated (Figure 3). The recovery of Sb was decreased with the amount of silica-gel. On the other hand, the recovery from precipitations was increased. The sum of the amount of Sb from the first digested solutions and those recovered from precipitations gave about 100 % for each amount of silica-gel.

Comparison of the Analytical Results of Common Heavy Metals

Contents of common heavy metals in these samples are shown in Table 1. These certified values were obtained by atomic absorption spectroscopy (8). With four common heavy metals, Ni, Cu, Cd and Pb, certified values and values observed by ICP-MS measurements were compared (Figure 4). Values measured by methods A and F were lower because of their low digestive and/or extractive ability, but others agreed approximately, with deviations from certified values in the following ranges; $\pm$10 % (Cd) and $\pm$25 % (Ni, Cu and Pb). It was considered that the errors increased with the amount of content because the results had to be calculated by extrapolating the calibration curve which was made by the same range of rare metals.

4 CONCLUSION

Based on the results of applying the six digestion methods to the measurement of rare metals in organic fertilizers using ICP-MS, the following can be concluded.

1) HNO_3-$HClO_4$-HF digestion was the most suitable digestion method for determination of Be, V, Ga, Sb, La and Bi by means of ICP-MS.

2) Sb tended to be adsorbed by uncertain mechanisms in the precipitations such as silicate, when the samples were digested by HNO_3-$HClO_4$ method. But Sb included in the precipitations could be recovered using HF digestion followed by acid dissolution.

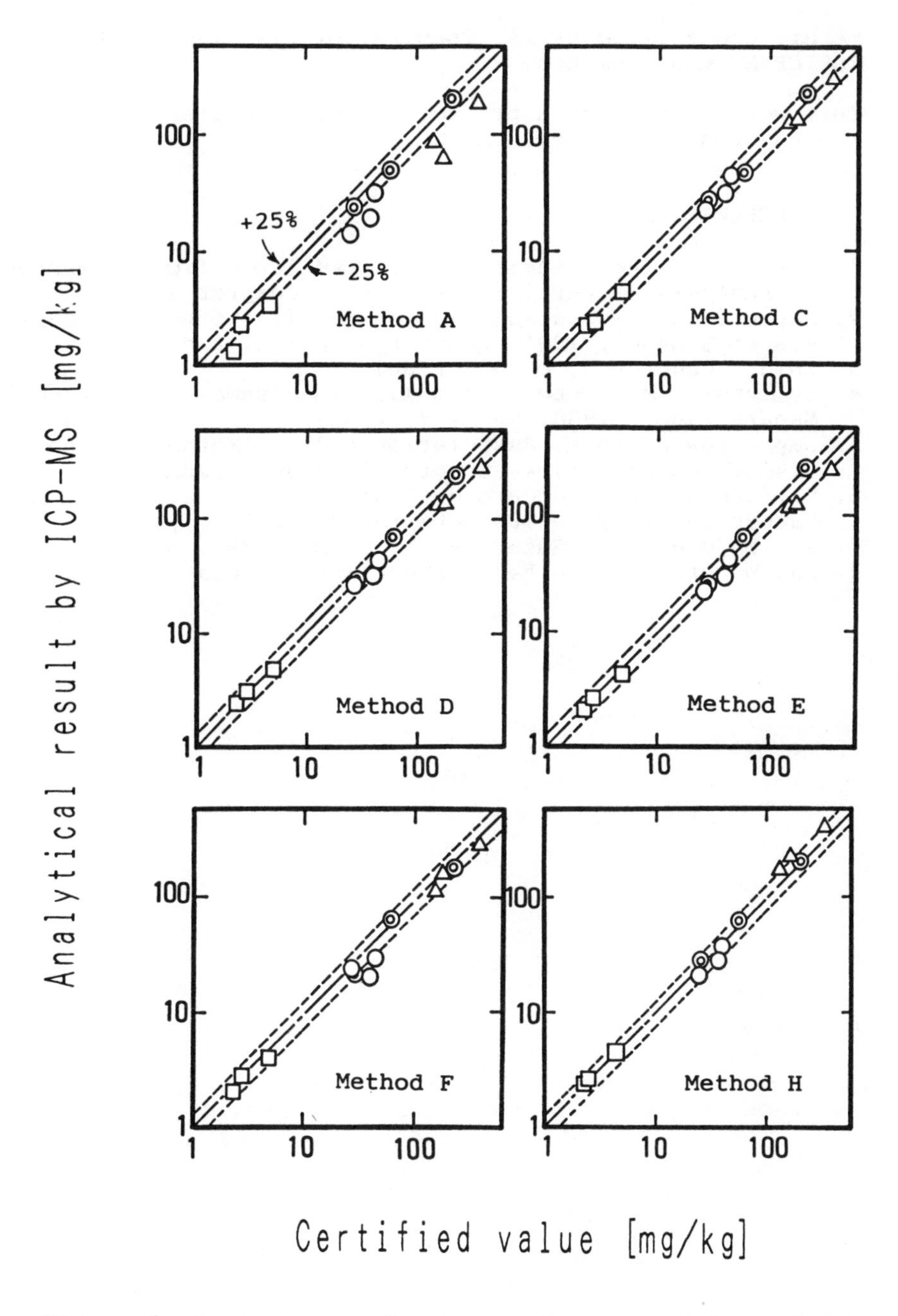

Figure 4 Comparison of common heavy metal contents [○ Ni; △ Cu; □ Cd; ◎ Pb] observed by ICP-MS with certified values.

ACKNOWLEDGMENT

We are grateful to Dr. A. Tsumura for his kind advice in ICP-MS measurements.

This work is part of a project funded by a grant from the Environment Agency, Japan.

REFERENCES

1. Natl. Inst. Agro-Environ. Sci. (ed.), 'Official Methods of Analysis of Fertilizers', 1987, Chapter 4, p.30. [E]
2. Japan Sewage Works Association (ed.), 'Methods of Analysis of Sewage', 1984, Chapter 5, p.302. [Jpn]
3. ibid., Chapter 5, p.302. [Jpn]
4. T., Asami, M., Kubota, and K., Minamisawa, J. Sci. Soil Manure, Jpn., 1988, 59, 197. [Jpn]
5. Japan Sewage Works Association (ed.), 'Methods of Analysis of Sewage', 1984, Chapter 5, p.303. [Jpn]
6. S., Yamasaki, 'Standard Methods of Soil Analysis', Hakuyusha, Tokyo, 1986, Chapter 28, p.171. [Jpn]
7. S., Bajo and U., Suter, Anal. Chem., 1982, 54, 49
8. Japan Fertilizer & Feed Inspection Association.

Determination of Boron in a Titanium Reference Material by ICP-MS

F. Vanhaecke, H. Vanhoe, C. Vandecasteele, and R. Dams

LABORATORY OF ANALYTICAL CHEMISTRY, UNIVERSITY OF GHENT, INSTITUTE OF NUCLEAR SCIENCES, PROEFTUINSTRAAT 86, B-9000 GHENT, BELGIUM

1 ABSTRACT

A method was developed for the determination of boron in titanium by inductively coupled plasma - mass spectrometry (ICP-MS). A commercially available teflon sample introduction system, leading to the desired low detection limits for boron, was used. The influence of the internal standard on the precision was studied and beryllium was selected as the internal standard. For the titanium analysed, BCR reference material 090, the ICP-MS result agrees with results obtained using other techniques. Several bars of the titanium reference material were supplied and a study of the homogeneity of boron in this material was carried out. Using analysis of variance on the results obtained for the different bars, the homogeneity of boron in the reference material could be estimated to be better than 2.1%.

2 KEY WORDS

inductively coupled plasma mass spectrometry; inert sample introduction system; internal standard; boron determination in titanium matrix; homogeneity study

3 INTRODUCTION

In its standard configuration an ICP-mass spectrometer is mainly intended for analysis of aqueous samples. Measurement of these samples involves the generation of a fine aerosol from the sample solution using a nebuliser and a spray chamber mounted within the ICP torch box. As a result, solid materials have to be taken into solution for analysis. For the dissolution of a number of solid samples hydrofluoric acid is needed. Using the standard glass sample introduction system, the determination of low boron concentrations is then impossible, since the hydrofluoric acid attacks the borosilicate glass, leading to very high blank values for boron. Evaporation of the hydrofluoric acid is not an efficient way to overcome this problem, since losses of boron will occur due to the volatility of borontrifluoride. Therefore, the use of a teflon sample introduction system for the determination of boron in samples taken into solution using hydrofluoric acid, e.g. titanium, is necessary.

This paper describes the determination of boron in titanium, BCR (Bureau Communautaire de Référence, Commission of the European Communities, Wetstraat 200, B-1049 Brussels, Belgium) CRM 090. As a result of the way in which the reference material was produced and the precautions taken in that production, the material was expected to be homogeneous. We checked the homogeneity of boron in this material by analysing different bars supplied by BCR. Before the determination, we carried out a study allowing to select an appropriate internal standard in order to obtain optimal precision and accuracy. To check the accuracy and the precision of the analytical procedure, the values obtained by ICP-MS were compared with the values obtained by other techniques.

4 EXPERIMENTAL

Instrumentation

The instrument used is a commercially available ICP-mass spectrometer, the VG PlasmaQuad (VG Elemental, Winsford UK). The preliminary study concerning the choice of the internal standard was carried out using the instrument in its standard configuration, with a torch of Fassel design, a Minipuls 2-Gilson peristaltic pump, a Meinhard concentric glass nebuliser (type Tr-30-A3) and a double pass Scott type spray chamber with surrounding liquid jacket, the temperature of which was controlled with a recirculating water refrigeration-heating system. Sampling cones (1.0 mm orifice) and skimmer cones (0.75 mm orifice) were made of nickel. For the measurements in this study we used a sample uptake rate of 0.9 ml min^{-1} and a nebuliser flow rate of 0.725 l min^{-1}, while the settings of the electrostatic lenses were optimised in order to obtain a maximum signal for ^{115}In. The other operation conditions used were identical as those used for the determination of boron in titanium (see table 1).

When HF-containing solutions are nebulised using the borosilicate glass sample introduction system, the blank value for boron is very high as a result of attack on the glass by hydrofluoric acid. Since hydrofluoric acid was used in the digestion procedure, a modification was necessary. A commercially available teflon sample introduction system (Inert Sample Introduction System ISIS,

VG Elemental, Winsford UK) was therefore used. With this type of sample introduction system a de Galan nebuliser is used and the water-cooled spray chamber and the connection elbow are made of teflon. The central tube of the torch is made of alumina. The solution never comes in contact with glass, so that the blank value for boron is low. In contrast to the glass sample introduction system, a second peristaltic pump (Minipuls 2-Gilson) is used with the teflon sample introduction system to ensure fluent drain, improving count stability. When this second peristaltic pump was set at a pumping rate of 5 ml min^{-1} the situation was found to be optimal. Also the sample uptake rate and the nebuliser gas flow rate were investigated, leading to the operation conditions summarized in table 1. The settings of the electrostatic lenses were optimised in order to obtain a maximum signal for the internal standard used, ^{9}Be.

<u>table 1</u> <u>PlasmaQuad operation conditions</u>

plasma r.f. power: forward: 1350 W
reflected: < 5 W
gas flows: plasma: 13.5 l min^{-1}
auxiliary: 0.65 l min^{-1}
nebuliser: 0.770 l min^{-1}
sample uptake rate: 1.7 ml min^{-1}
spray chamber temperature: 10 °C
ion sampling depth: 10 mm (from load coil)
vacuum: expansion stage: 2.5 mbar
intermediate stage: 1.0 10^{-4} mbar
analyser stage: 5.4 10^{-6} mbar
resolution: low mass: 3.50
coarse: 0.3
fine: 5.20

Reagents and Solutions

For the preliminary study concerning the choice of the internal standard a multi-element solution, containing a number of elements, covering the total mass range and possibly being suitable to be used as internal standard, was prepared. The elements were present at a concentration of 100 μg l^{-1}. This multi-element solution was obtained by dilution of commercial standard solutions (1 g l^{-1}) with 0.14 M HNO_3.

As the response of ICP-MS is known to be linear over a very wide range, only one standard solution (prepared at least in triple), with a similar boron concentration as the sample solutions, was used for calibration. Every boron standard solution was obtained by successive dilution with 0.14 M HNO_3 of more concentrated boron solutions prepared by dissolving boric acid H_3BO_3 (CBNM IRM 011) in 0.14 M HNO_3. The 0.14 M nitric acid used was obtained by hundred-fold dilution of concentrated nitric acid, purified by subboiling distillation, with double distilled water. An 0.14 M HNO_3 solution was used as a blank for the standard solutions. Both the standard solutions and the blank contained 100 $\mu g\ l^{-1}$ beryllium as an internal standard. Thoroughly precleaned polyethylene volumetric flasks and glass pipettes were used throughout.

For the digestion procedure analytical-reagent grade hydrofluoric acid (50%, RPL-UCB, Belgium) was used. The titanium samples were supplied as cubes with an average weight of about 80 mg. Sample solutions were prepared in the following way: a cube was accurately weighed and brought into a thoroughly precleaned teflon vessel. First 5 ml of double distilled water was added. Next 700 μl of hydrofluoric acid was added in several portions. The mixture was left until all of the titanium had dissolved. The solution obtained was transferred quantitatively in a polyethylene volumetric flask, and diluted to a final volume of 50 ml using 0.14 M nitric acid. In the course of the dilution, beryllium was added to obtain a final concentration of 100 $\mu g\ l^{-1}$. The same procedure without sample was used to obtain blank solutions for the samples.

Measurements

The measurements concerning the internal standard study were made using the mass scanning mode of data acquisition (120 scan sweeps, 320 μs dwell time per channel, 4096 channels, mass range 4-255 u, acquisition time = ± 157 s). For investigating the short term stability 10 successive measurements were made. The total analysis time was ca. 30 min in this case. For investigating the long term stability the multi-element solution was measured every quarter of an hour during a 4 h period resulting in 17 measurements. These measurements were repeated on different days.

The analyses were performed using the mass scanning mode of data acquisition. The scan conditions are summarized in table 2.

Table 2 Data acquisition-Quantitative scanning

number of scan sweeps:	120
dwell time per channel:	320 μs
number of channels used:	512
mass range:	8-13 u
acquisition time:	ca. 20 s

For the determination of boron in the titanium reference material, the following procedure was used. For every titanium bar supplied two samples were analysed for boron. Since eight bars were to be analysed, we obtained a total of sixteen sample solutions. Two blank solutions for the standards and four blank solutions for the samples were prepared. The boron standard solutions contained about 100 $\mu g\ l^{-1}$ boron. To limit the total analysis time, we divided the solutions into two batches, each containing a blank solution for the standards, two blank solutions for the samples, eight samples, one for each bar, and the standard solutions. On each of the two batches two series of measurements were performed on different days. During each series, the blank for the standards was measured first, followed by the blank for the samples. Then samples and standards were measured alternately. Each blank, sample or standard solution was measured three times. After these three measurements performed on a sample or standard solution, the sample introduction system was rinsed for a number of minutes using 0.14 M nitric acid in order to avoid possible memory effects.

5 CALCULATIONS

The peak integrations were perfomed by the software using the constant range integration mode. In this mode every signal obtained is the result of peak area integration over 0.8 u around the nominal mass.

For the preliminary experiment concerning the choice of the internal standard we used the software to integrate the peaks, to calculate signal ratios and the standard deviations and relative standard deviations (RSDs) of the results for the signals of the different elements and the calculated signal ratios. To evaluate the use of an internal standard the following procedure was used: we "selected" one element while the other elements were considered, one after the other, as the internal standard. For each internal standard the ratio ("selected" element/internal standard) was calculated for every measurement. The standard deviation and relative standard deviation for the 10 or 17 signal ratios, respectively for the short and long term stability experiment, were calculated. This was repeated for every element chosen as internal standard. The obtained RSDs could then be compared: firstly they were compared with the RSD for the 10 or 17 results for the signal of the selected element to check if the use of an internal standard gives an improvement and secondly they were compared with one another to be able to choose the most advantageous internal standard.

For the determination of boron in titanium ^{11}B seems the obvious choice as the nuclide to be monitored as a result of its higher isotopic abundance (ca. 80% for ^{11}B versus ca. 20% for ^{10}B). The signal ratios ($^{10}B/^{9}Be$) and ($^{11}B/^{9}Be$) were supplied automatically and the mean, the standard deviation and the relative standard deviation for the results of the three measurements were calculated for each blank, sample or standard solution.

For the calculation of the boron concentration, for every sample solution the blank-subtracted value of the ratio ($^{11}B/^{9}Be$) was divided by the blank-subtracted value of the ratio ($^{11}B/^{9}Be$) for the standard and multiplied by the concentration of the standard. In this way the boron concentration in the sample solutions was obtained and the boron concentration in the solid material could be calculated. Since eight different bars were supplied for each of which two samples were analysed this gave us sixteen results, each result being the mean of the results of 2 different days, each of which is based on 3 measurements on the sample. The mean and the standard deviation of these sixteen results were then calculated to obtain a final result for the concentration of boron in titanium.

6 RESULTS AND DISCUSSION

Choice of internal standard

The preliminary internal standard experiments enabled us to reach some general guidelines concerning the choice of an internal standard when using the mass scanning mode of data acquisition. Fig.1 gives the results for a long term stability measurement with yttrium as the "selected" element. This graph represents a situation which is far from optimal, since the relative standard deviation for the 17 results of the yttrium signal itself, obtained during the four hour investigation period, was ca. 8% where normally relative standard deviations are smaller than 4% over such a period. As can be seen from the graph, a smaller relative standard deviation (down to ca. 3%) resulting in more precise results can be obtained using an internal standard. However, when the stability of the ion signal itself is better, e.g. RSD less than 4%, the improvement obtained by the use of an appropriate internal standard is more limited, and in cases where optimal stability of the ion signal is established practically no further improvement is obtained using an internal standard. When the mass scanning mode of data acquisition is used, the mass number of the internal standard seems to be of great importance. As a general trend fig.1 shows that the closer the mass number of the internal standard is to the mass number of the selected element, the greater the improvement. As the difference in mass number between the selected element and the internal standard increases, the improvement is more limited. This graph also shows that when the difference in mass number between the selected element and the internal standard becomes too large, the relative standard deviation for the signal ratio (selected element/internal standard) can even exceed the relative standard deviation for the results of the ion signal for the selected element itself.

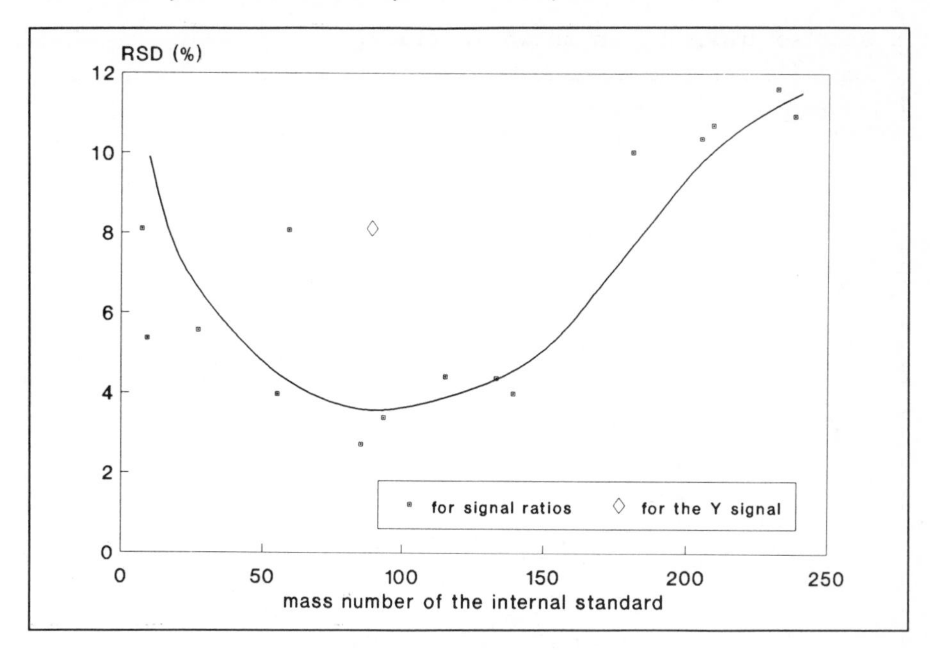

Figure 1: Relative standard deviations (%) for the signal of a "selected element" and for ratios ("selected" element / internal standard) as a function of the mass number of the internal standard chosen. Results presented are for a long term stability experiment with yttrium as the "selected" element.

Boron also follows this general trend, since for the same long term stability experiment as in fig. 1 we found an RSD of 9.8% for the ^{11}B-signals, while for the RSDs for the signal ratios ($^{11}B/^{9}Be$), ($^{11}B/^{115}In$) and ($^{11}B/^{205}Tl$) respectively 5.1%, 8.1% and 13.3% were found. These three ratios are of particular interest, since we tend to use ^{9}Be, ^{115}In and ^{205}Tl as internal standard respectively in the low-mass, middle-mass and high-mass region. Analogous results are obtained for the short term stability study: as RSD for the ^{11}B-signals 5.1% was found, while for the signal ratios ($^{11}B/^{9}Be$), ($^{11}B/^{115}In$) and ($^{11}B/^{205}Tl$) respectively 2.3%, 3.6% and 5.6% were found. Again this study was carried out under non-optimal conditions, since normally an RSD smaller than 2% is obtained for such short term stability experiments.

The importance of the mass number of the internal standard looks surprising at first sight, since we would rather expect the first ionisation energy of the internal standard to be the determining factor. Plotting the relative standard deviations for the signal ratios as a function of the first ionisation energy however results in a scatter diagram, as shown in fig.2.

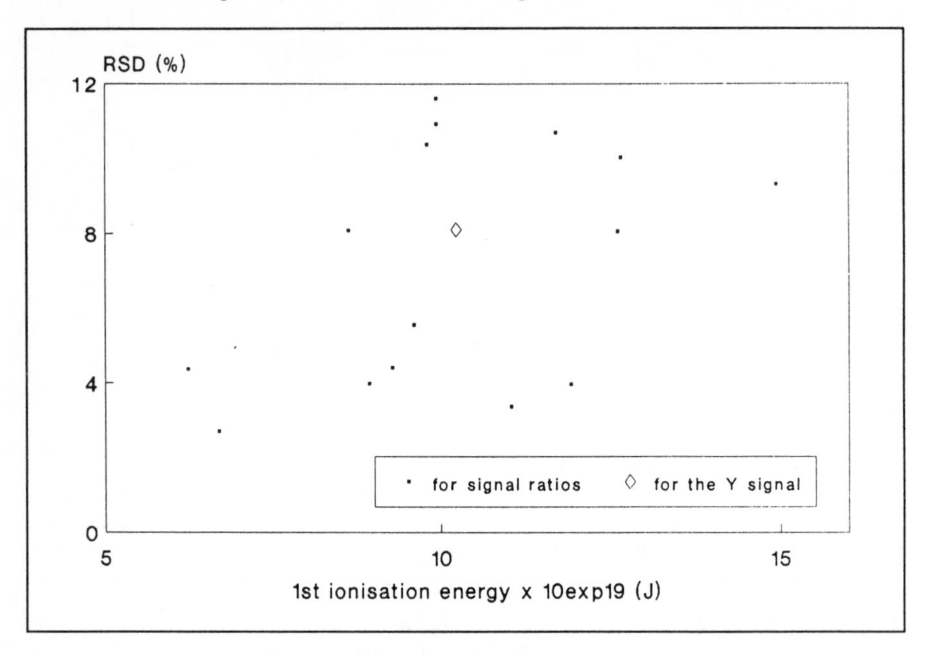

Figure 2: Relative standard deviations (%) for the signal of a "selected" element and for ratios ("selected" element / internal standard) as a function of the first ionisation energy of the internal standard chosen. Results presented are for a long term stability experiment with yttrium as the "selected" element.

Further experiments are needed to explain the observed behaviour. To obtain optimal precision an internal standard as close in mass number as possible to the analyte elements should be selected.

This choice also has an additional advantage, since it is our experience that also signal suppression can be mass-dependent. In some cases we established that the extent to which the ion signal is suppressed (or enhanced) depends on the mass number of the nuclide. This is shown in fig.3, in which for several elements present in a 0.5 M sulphuric acid matrix the residual signal (relative to the signal in a 0.14 M HNO_3 matrix) is plotted. As can be seen from this graph, the signal is suppressed to 56% from its value in 0.14 M HNO_3 for scandium and to 77% for uranium. To obtain accurate results in such cases, the internal standard should be chosen as close in mass number as possible to the mass number of the analyte elements. However, there is still much discussion about the mass dependency of non-spectroscopic interferences[1,2,3] and further investigation is being carried out to obtain more insight in this matter.

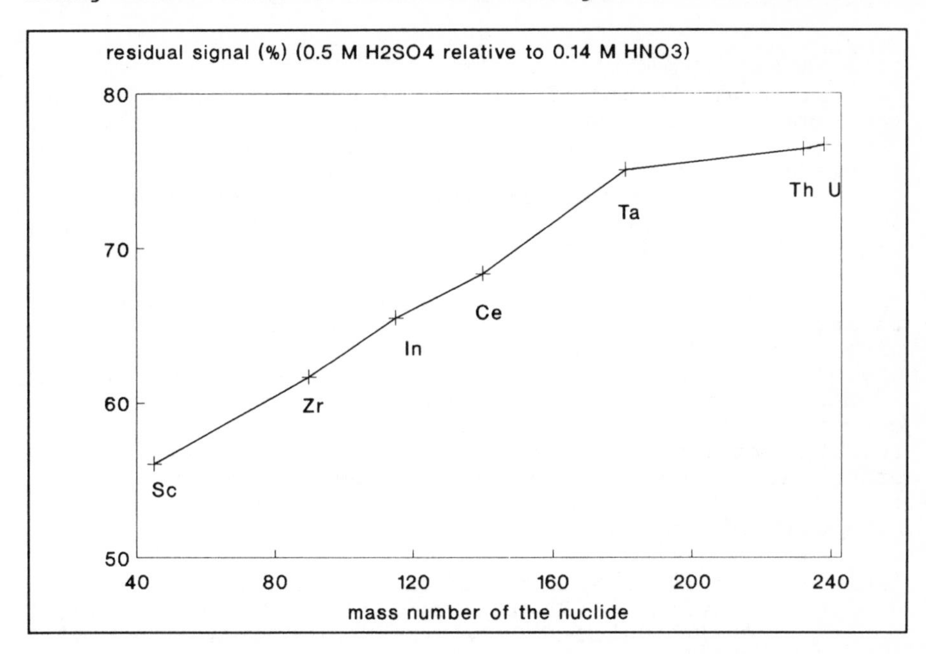

Figure 3: The residual signal (relative to the signal in a 0.14 M HNO_3 matrix) for several elements present in a 0.5 M H_2SO_4 matrix as a function of the mass number of the nuclide.

For the determination of boron in titanium signal suppression ranged from ca. 5% to ca. 20%. Beryllium was chosen as the internal standard.

Detection limit and blank

With the teflon sample introduction system we were able to reach detection limits for boron below 1 $\mu g\ l^{-1}$ (ca. 0.4 $\mu g\ l^{-1}$). This detection limit was calculated on the basis of the standard deviation of 10 successive measurements of the blank solution, using the 3s-criterion. The blank value for boron however, differed from day to day, and its origin is not quite clear. For the blank value for boron in 0.35 M HF, we calculated on the basis of the ^{11}B nuclide a mean value of 1.7 $\mu g\ l^{-1}$ with a standard deviation of 1.2 $\mu g\ l^{-1}$ based on the results of 10 determinations performed on different days. Moreover this residual blank value was comparable with the blank value for an 0.14 M HNO_3 solution. Using the glass introduction system blank values for boron range up to several hundreds $\mu g\ l^{-1}$ in HF-containing solutions. For an 0.14 M HNO_3 solution blank values for boron were comparable using the glass and the teflon sample introduction system.

Results

First we checked if the sample and the standard showed the same isotopic composition. No significant difference (t-test) was found between the ($^{10}B/^{11}B$) ratios for the sample and standard solutions. This ratio differed however from the certified isotopic ratio of the H_3BO_3 standard as a result of mass fractionation, which is particularly important in the low mass region.

The determination of the boron concentration in titanium resulted in a mean concentration of 28.98 $\mu g\ g^{-1}$ with a standard deviation of 0.85 $\mu g\ g^{-1}$. Other techniques used for this determination were inductively coupled plasma - atomic emission spectrometry, spectrophotometry and charged particle activation analysis. Comparison of our result, obtained by ICP-MS, with those obtained by the other techniques shows a good agreement, as can be seen from table 3. Also the standard deviation is comparable with that of other techniques.

Table 3 Results for the determination of boron in the titanium reference material obtained by various analytical techniques. The ICP-MS value is the result of this work.

Technique (n° of samples)	mean concentration (in $\mu g\ g^{-1}$)	standard deviation (in $\mu g\ g^{-1}$)
ICP-AES (4)	30.2	2.1
ICP-AES (4)	26.8	2.6
ICP-AES (4)	30.50	0.58
ICP-AES (4)	26.0	1.1
SPEC (4)	27.92	0.50
SPEC (8)	26.8	1.0
CPAA (4)	28.72	0.61
ICP-MS (16)	28.98	0.85

In order to investigate the homogeneity of boron in the titanium reference material, one-way analysis of variance[4] (ANOVA) was carried out on the results obtained from each of the two batches. For such a batch, the variation between the results obtained for the different bars can be attributed to two possible sources. The first source of variation, which is of course always present, is due to random error. The second possible source of variation is due to what is known as controlled or fixed-effect factors, amongst which in this case inhomogeneous distribution of boron in the titanium bars. ANOVA is a statistical technique which can be used to separate and estimate different causes of variation. In this case, it enabled us to make a distinction between the variation due to the random error in the measurements and the variation due to the inhomogeneity of boron in the different titanium bars, together with other random-effect and fixed-effect factors, mainly dissolution errors. As a result we were able to estimate an upper limit for the inhomogeneity of boron in the titanium bars. A summary of the formulae used in one-way ANOVA is given in table 4a. Table 4b presents a summary of the results of one-way ANOVA carried out on the two batches. At first the between-sample and the within-sample mean squares were calculated. A one-tailed F-test (P=0.05) indicated that for each of the two batches the between-sample mean square is significantly greater than the within-sample mean square. This indicates that the null hypothesis (all the samples are drawn from the same population with mean μ and variance σ_0^2) has to be rejected, the sample means do differ significantly. The within-sample mean-square can now be used as an estimation of σ_0^2, the variance attributed to the random error in the measurements. The between-sample mean square on the other hand gives an estimation of $(\sigma_0^2+n\sigma_1^2)$ allowing to calculate σ_1^2, the variance due to the inhomogeneity of boron in the titanium bars together with some other random and fixed-effect factors such as dissolution errors, assuming normal distribution. Dividing σ_1 by the mean concentration for the batch and expressing the result as a percentage we could estimate the homogeneity of boron in the titanium reference material to be better than 2.1%.

Taking into account the results of the different laboratories and the result of the homogeneity study, the BCR proposed for certification a boron concentration of 28.2 ± 1.7 $\mu g\ g^{-1}$ in the titanium reference material CRM 090.

F. Vanhaecke is a Research Assistant of the Belgian National Fund for Scientific Research.
C. Vandecasteele is a Research Director of the Belgian National Fund for Scientific Research.

Table 4a Summary of the formulae used in one-way ANOVA.

Source of variation	Sum of squares	Degrees of freedom	Mean square
Between-sample	$\Sigma T_i^2/n - T^2/N$	h-1	Sum of squares / degrees of freedom
Within-sample	by subtraction	by subtraction	
Total	$\Sigma_i \Sigma_j x_{ij}^2 - T^2/N$	N-1	

In which: the number of samples = h
the number of measurements per sample = n
the total number of measurements = N = nh
the sum of the measurements on the ith sample = T_i
the sum of all the measurements, grand total = T
the jth measurement on the ith sample = x_{ij}

The mean squares are tested on significant difference using a one-tailed F-test.
If a significant difference is established:
- the within-sample mean square = estimation of σ_0^2
- the between-sample mean square = estimation of $(\sigma_0^2 + n\sigma_1^2)$

Table 4b Summary of the results of one-way ANOVA carried out on the results (expressed in $\mu g\ g^{-1}$) of the two batches.

	Source of variation	Sum of squares	Degrees of freedom	Mean square
Batch 1	Between-sample	20.73	7	2.9608
	Within-sample	24.59	40	0.6146
	Total	45.31	47	
Batch 2	Between-sample	16.50	7	2.3577
	Within-sample	21.64	40	0.5408
	Total	38.14	47	

Batch 1: F = 4.82, $\sigma_0^2 = (0.7840)^2$, $\sigma_1^2 = (0.6253)^2$

Batch 2: F = 4.36, $\sigma_0^2 = (0.7354)^2$, $\sigma_1^2 = (0.5503)^2$

[Critical value of F $F_{7,40}$ (P=0.05) = 2.25]

REFERENCES

1. J.J. Thompson and R.S. Houk, Appl. Spectrosc., 1987, 41, 801.
2. G.R. Gilson, D.J. Douglas, J.E. Fulford, K.W. Halligan and S.D. Tanner, Anal. Chem., 1988, 60, 1472.
3. C. Vandecasteele, M. Nagels, H. Vanhoe and R.Dams, Anal.Chim. Acta, 1988, 211, 91.
4. J.C. Miller and J.N. Miller, 'Statistics for Analytical Chemistry', Ellis Horwood Limited, Chichester, 1984, Chapter 3, p.67.

Applications of ICP-MS with Sample Introducton by Electrothermal Vaporization and Flow Injection Techniques

Uwe Voellkopf[1], Andreas Guensel[1], Michael Paul[1], and Helmut Wiesmann[2]

[1] BODENSEEWERK, PERKIN-ELMER GMBH, POSTFACH 101164, 7770 UEBERLINGEN, GERMANY
[2] SPECTROTEC GMBH, WEIDENSTRASSE 18, 6097 TREBUR 1, GERMANY

1 INTRODUCTION

Although a young analytical technique, inductively coupled plasma mass spectrometry (ICP-MS) has become widely accepted as a powerful tool for trace and ultra-trace elemental and isotopic analysis. ICP-MS, like inductively coupled plasma atomic emission spectrometry (ICP-AES) or graphite furnace atomic absorption spectrometry (GF-AAS), was initially developed and is still predominantly used for the analysis of solutions. However, ICP-MS is also well suited to be used as a detector for other sample introduction devices such as high pressure liquid chromatography [1 - 3], laser sampling [4, 5], flow injection techniques [6 - 11] and electrothermal vaporization [12 - 21].

At present the most frequently used alternate sample introduction technique for ICP-MS is the direct analysis of solid samples using a laser sampling accessory. This technique is extending the applicability of ICP-MS to a wide variety of new sample types that have previously been inaccessible due to the practical limitations of conventional dissolution techniques.

Flow injection analysis (FIA) is a rapidly growing analytical technique which has been applied to a number of standard methods. Since its introduction by Ruzicka and Hansen [22] in 1975, FIA has been applied in many different analytical fields and is an established technique covering a wide range of applications with over 1000 publications [23]. FIA has been used successfully for ICP-MS [5 - 11] as an alternative sample introduction technique for different applications. Flow injection provides a number of advantages over conventional procedures of solution handling in atomic absorption spectrometry (AAS), ICP atomic emission spectrometry (ICP-AES) and ICP mass spectrometry (ICP-MS). These include high efficiency, sample and reagent consumptions in the sub-milliliter range, a high tolerance of dissolved solids, automatic online matrix separation and/or analyte pre-concentration, automatic dilution and addition of, e.g., internal standards along with a high degree of automation. We have used a commercially available computer controlled flow injection system with a hydride generation accessory for the ICP-MS multielement determination of hydride forming elements such as As, Se, Sb, Hg and Bi. Initially instrumental and analytical parameters were optimized starting from the recommended analytical conditions given by the manufacturer of the flow injection system for flow injection hydride generation AAS. Using optimized conditions for the ICP-MS, some interference studies were performed, in particular, the interference of some transition elements on the efficiency of the hydride generation process. Finally, the method was applied to the analyses of reference samples (NIST 1643b water and NRC sea water).

Electrothermal vaporization [ETV] is an analytical technique which requires only very small sample volumes and overcomes some of the limitations of solution nebulization. The technique of electrothermal vaporization was first described as an alternate sample introduction technique for ICP-AES in 1974 [24]. Electrothermal vaporization can provide a number of significant advantages compared to ICP-MS analysis with conventional solution nebulization. These advantages are: (*i*) the use of micro sample volumes (typically 5μL to 100μL), (*ii*) a higher transport efficiency with respect to solution nebulization, (*iii*) the ability to analyze organic samples and samples containing high concentrations of dissolved solids and solid samples, (*iv*) separation of analyte elements from the matrix by applying special furnace temperature/time programs and chemical modification techniques, (*v*) the potential to eliminate metal-oxide and matrix -induced polyatomic interferences and (v*i*) extremely low absolute detection limits. For these reasons ETV-ICP-MS is of great advantage whenever very small sample volumes must be analyzed for multiple elements, or when conventional ICP-MS analysis is handicapped by metal-oxide or polyatomic interferences.

A number of different ETV configurations have been discussed in the literature[12 - 21], such as metal filament and graphite rod atomizers enclosed in quartz domes. Results of experiments with a modified commercial heated graphite furnace atomizer (HGA) for ETV-ICP-AES analyses were reported in 1982 [25] by Crabi and co-workers. Recently, another paper describing the use of a slightly modified commercial heated graphite furnace atomizer has been published by Shen and co-workers[21]. The modifications made included: (*i*) the use of a graphite tube without sample introduction port, (*ii*) sample introduction by means of a tungsten rod inserted into the graphite tube from one end of the furnace work head through a special insertion port which replaced the standard quartz glass window, (*iii*) sample outlet through the opposite tube end with an extractor tube which replaced the second quartz glass window and (*iv*) an additional gas inlet at one tube end via a hole drilled through one cooling cylinder and one of the graphite contacts for the introduction of a coolant gas.

The approach of our recent hardware development project was to design a simple ETV modification kit for commercial Perkin-Elmer heated graphite furnace atomizers (HGA). The aim was to make the modification kit very easy to retrofit to existing HGAs in the field, such that a quick change between the use of the HGA as graphite furnace atomizer (AAS) and electrothermal vaporizer (ICP-AES and ICP-MS) would be póssible. In addition, it was considered essential that the ETV system should be compatible with the standard HGA autosampling system, slurry sampler and the cup-in-tube technique [26 - 27] for the direct analysis of solid samples. Our experience with an earlier experimental setup [17] served as a useful basis for the design and performance specifications.

In the first stage of performance evaluation of the ETV-ICP-MS system, sensitivities, signal-to-noise ratios and detection limits were determined for a number of elements across the mass range from Li to Bi. These studies were performed in both single ion monitoring and multielement modes. Finally, the potential of ETV-ICP-MS for performing very fast screening analysis was evaluated.

2 EXPERIMENTAL

ICP-MS Instrumentation

The ICP mass spectrometer used for this work was a standard Perkin-Elmer Sciex ELAN 5000 (Perkin-Elmer Sciex, Norwalk, CT, USA).

Single Ion Monitoring and Multichannel Analysis Conditions. For single element analysis, the ELAN 5000 was operated in its single ion monitoring mode. For multielement determinations the ELAN 5000 was operated in its multichannel mode, which allows multiple scans per second across the mass spectrum from mass 6 (Li) to mass 238 (U). For most experiments the dwell times were adjusted such that a sufficient number of sweeps per second could be performed for the number of elements (masses) selected in the method. As a rule of thumb, one can assume that in GF AAS the analyte absorption signal is measured once every 20 milliseconds. Thus, when operating the ICP mass spectrometer in single ion monitoring, dwell times of approximately 20 milliseconds are more than sufficient for the exact measurement of fast ETV transient signals. However, when multiple elements are to be determined, then the total number of elements selected should be measured within 20 milliseconds. Assuming that 5 elements are to be determined, a dwell time of approximately 4 milliseconds should be used. The larger the number of elements, the shorter the dwell time must be.

In this mode of operation, where the total mass spectrum can be scanned many times per second, a large number of isotope-intensity data points must be handled by the computer. In order to avoid an overflow of data at the external computer, the internal ELAN controller has a built-in data buffer which allows the intensity data to be collected at a very high speed before transfer to the external controller at the end of the measurement.

The transient signals produced by flow injection sample introduction systems are much slower than those from ETV. In addition the user may influence the signal peak shape by proper selection of the sample introduction loop volume.

ELAN ICP-MS Conditions. Standard plasma parameters were used in all experiments. The ion optics settings need not be re-adjusted when switching from sample introduction by conventional solution aspiration to hydride generation flow injection or electrothermal vaporization. The performance of the spectrometer was quickly assessed once every morning using a mixed standard solution of Li, Rh and Pb (10 ng/L) and the intensities for the three elements were examined on the computer screen utilizing the real time intensity/time graphical display. Upon completing this early morning setup test, the spray chamber was removed from its quick change torch mount and replaced by the transfer tube of either the FIAS-200 system or the electrothermal vaporizer.

Flow Injection Accessory

The Flow injection system used for the work presented in this paper was a standard FIAS-200 with AS-90 autosampler (Bodenseewerk, Perkin-Elmer, GmbH, Ueberlingen, FRG). The FIAS-200 is a high performance computer controlled flow injection accessory. It consists of two peristaltic pumps, which may be controlled independently from each other, for the transport of carrier, reagent and sample flows, a high precision flow injection valve, a chemifold and a random access autosampler. The FIAS-200 may be controlled either directly through the spectrometer control software or by means of special stand alone software which was developed for the use of the FIAS-200 with older atomic absorption spectrometers. No modification was necessary to the FIAS-200 equipment prior to its connection to the ELAN 5000 ICP mass spectrometer.

Electrothermal Vaporization Accessory

The ETV device used at this stage of hardware and method development work was a modified HGA-400 with an AS-40 autosampler (Bodenseewerk, Perkin-Elmer, GmbH, Ueberlingen, FRG). The modification kit has been developed in a joint project between Spectrotec GmbH (Trebur 1, FRG) and Bodenseewerk, Perkin-Elmer GmbH. The modifications to the work head of the standard graphite furnace are very simple. A T-piece is inserted into one of the two internal gas tubes. Thus, the carrier gas (re-directed nebulizer gas) could be inserted in parallel with one of the internal gas flows. Further, the quartz glass window at the oposite site of the carrier gas inlet was replaced with a ceramic tube (inner diameter approx. 0.6mm). This ceramic tube was connected to a simple Tygon tube through which the vaporized sample was transported to the torch of the ELAN ICP mass spectrometer. Finally, a double metal ring was constructed around the graphite contacts of the furnace. This modification allows the autosampler or the user to pipet the sample into the graphite tube. Then the furnace temperature/time program is initiated. By means of the furnace programmer remote controls, the outer ring is then rotated shortly before the vaporization step. This automatically closes the sample introduction port of the graphite furnace. During the vaporization step, the external argon sheath gas of the furnace work head forces the vaporized material to leave the graphite tube through the ceramic extractor tube and thus prevents, to a large extent, the loss of vaporized material through the sample introduction port of the graphite tube as would be normal with an open furnace. This simple modification kit can be easily retrofitted to conventional Perkin-Elmer heated graphite furnace atomizers (HGA-400, HGA-500 and HGA-600 series). The conversion in both directions can be performed within less then 30 minutes and there is no requirement to drill any holes in the furnace workhead. One of the major advantages of using a standard graphite furnace atomizer as an ETV system is that a standard autosampler, normal graphite tubes and any other standard HGA accessories can be used directly along with the application of many years of experience with HGA type graphite furnaces.

Samples and Reagents

Standard solutions were prepared by diluting a 23 element ICP-MS multielement standard (supplier: Merck, Darmstadt, FRG; ICP Multielement Standard IV) with deionized water (18MOhm/cm). This standard has been used with great success in our laboratory during the last two years. All elements are present as nitrates at a concentration of 1000 mg/L in the stock solution. No sample prepararation was required for the certified reference materials NIST 1643b water, NRC NASS-1 sea water and NRC SLRS-1 Riverine water.

Optimization of ETV Temperature/Time Programs

In GF-AAS the optimum thermal pretreatment and atomization conditions differ greatly between elements, therefore in ETV-ICP-MS a dependence of the signal/noise ratios on the ETV settings is also to be expected. Thermal pretreatment and vaporization studies were performed for several elements of different volatility operating the ELAN 5000 in its single ion monitoring mode. Our expectations that optimum GF-AAS and ETV-ICP-MS conditions should not differ greatly were confirmed.

For multielement analysis compromise conditions had to be selected for the ETV temperature/time program (Table 1). The thermal pretreatment temperature was selected such that losses of the most volatile element (Cd) during the pretreatment step was avoided. The vaporization temperature was selected such that even the most refractory elements (Rh and Ni) could be volatilized from the wall of the pyrolytically coated tube with sufficient sensitivity.

Table 1: A compromise ETV temperature/time program for multielement analysis.

Step	1	2	3	4	5	6	7
Temp. (°C)	90	120	250	250	2600	20	2650
Ramp Time	5	10	5	1	1	1	1
Hold Time	10	15	10	10	6	10	6
Close				XX	XX	XX	
Read					XX		

Optimization of Hydride Generation FIA Parameters

The successful application of hydride generation FI-ICP-MS requires the careful optimization of several instrumental and analytical parameters. It is well known from AAS hydride generation, that the optimum conditions for achieving the best signal/noise ratios and maximum freedom from interferences differ from element to element. Thus ICP-MS multielement determinations are only possible using compromise conditions and parameters for the $NaBH_4$ concentration, the reaction loop volume, injection loop volume and the FIAS-200 pump speeds must be adjusted.

$NaBH_4$ Concentration. The effect of different $NaBH_4$ concentrations on the analytical signal of As was studied. A rise in the As signal intensities was found with increasing $NaBH_4$ concentrations. However, the baseline (blank) signal intensity was rising as well. This was due to As contamination of the $NaBH_4$ used. Due to these contamination problems, the concentration of $NaBH_4$ must be kept as low as possible and a concentration of 1g/L to 2g/L in a solution containing 0.4% NaOH was found to be a good compromise.

Reaction Loop Volume. The reaction loop of a flow injection hydride generation device allows for the proper mixing of the reagent ($NaBH_4$) with the sample, the carrier solution (HCl or HNO_3) and the argon transport gas. The influence of reaction loop lengths between 26cm and 260cm on the analytical signals for As in both, acidified standard and matrix solutions (containing high concentrations of Cu and Ni) were studied. For aqueous solutions it was found that as the reagent loop length was increased the analytical signals initially decreased (52cm to 104cm) but at very long loop lengths (104cm to 260cm) rose again. This behaviour is due to two effects occurring in parallel, a) dilution of the sample with increasing tube length and b) increasing reaction time causing improvements in sensitivity. However, since the use of very long reaction loop lengths did not result in better sensitivities than those obtained with very short reaction loops of less than 50cm, it was decided to use short reaction loops. Short reaction loops are also an advantage when determining hydride forming elements in samples containing high concentrations of transition elements such as Ni or Cu. The interference from transition elements on the the hydride forming elements increases with the reaction loop length leading to losses in the analyte signals.

Injection Loop Volume. With the FIAS-200 the minimum injection loop volume is approximately 40µL. In principle there is no maximum injection loop volume since users may easily construct injection loops of any length. However, typical injection loop volumes are in the range of 100µL to 1000µL. With injection loop volumes in the range of 40µL to approximately 700µL, relatively sharp transient signals were found. With injection volumes of more than 700µL, transient signals with plateaus at the peak maximum were found. For practical reasons it was decided to perform all further work presented in this study with an injection loop volume of 500µL. All important mass spectrometer and FIAS-200 parameters are summarized in Table 2.

Table 2: ELAN and FIAS-200 parameters for multielement hydride generation FI-ICP-MS

Parameter	Value	Parameter	Value
Sweeps/Replicate:	variable	Auxilliary Gas :	0.9 L/min
Dwell Time:	20 ms	Pump 1 (Sample):	120 rpm
RF Power:	1100 W	Pump 2 (HNO_3 and $NaBH_4$):	100 rpm
Plasma Gas:	14 L/min	Injection Loop:	500 µL
Support Gas:	1 L/min	Reaction Loop:	200 µL
		Carrier Gas:	0.45 L/min

3 RESULTS AND DISCUSSION

Determination of Hydride Forming Elements by FI-ICP-MS

Sensitivity and Detection Limits. Examples for both, the sensitivity and reproducibility which can be achieved in the multielement analysis of hydride forming elements by FI-ICP-MS are shown in Figure 1. By far the highest sensitivity was achieved for Sb (approximately 120000 counts/second for 1ng/mL). For Bi, Hg and As sensitivities between 50000 counts/second and 18000 counts/second per 1ng/mL were measured. The lowest sensitivity (300 counts/second per 1ng/mL) was achieved for Se using the isotope at 82 dalton. The reproducibility for all elements studied simultaneously (multielement measurement) was between 3% and 5% RSD (5 replicates).

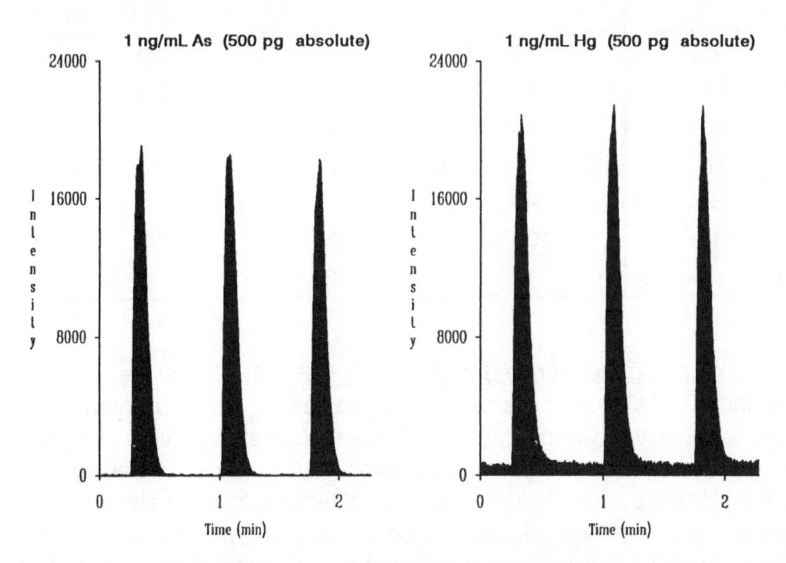

Figure 1: Typical hydride generation FI-ICP-MS transient signals for As and Hg (1ng/mL, injection loop volume 0.5mL)

The results presented in Figure 1 as well as the detection limits summarized in Table 3 demonstrate the excellent detection power of FI-ICP-MS for the analysis of hydride forming elements and Hg. The sensitivity for Hg can be further improved by combining the cold vapor amalgamation technique with FI-ICP-MS. The sensitivity with only a single collection/heat out step (0.5mL injection volume) is increased approximately 4 times over conventional FI-ICP-MS Hg determinations. Further improvements in the detection power could be achieved by multiple amalgamation steps prior to the heat out of the Au/Pt net.

The major problem observed during the amalgamation experiments was due to the Hg contamination in the diluent, acid and reagent solutions at these extremely low ppt concentration levels. Even in the argon plasma gas sub ppt concentrations of Hg were found. This contamination sets the practical limits for the Hg determination by cold vapor amalgamation FI-ICP-MS.

Table 3: Relative detection limits of hydride generation FI-ICP-MS and cold vapor amalgamation FI-ICP-MS (Hg) in comparison to solution aspiration ICP-MS and hydride generation FI-AAS; all values are based on an injection loop volume of 0.5mL and are given in ng/L

Element	FI-AAS	FI-ICP-MS	ICP-MS (conv. nebulization)
As	50	5	200
Se	40	60	500
Sb	45	3	1
Te	110	8	10
Bi	40	10	1
Hg	100	15	100
Hg (Amalgam)	12	3	100

Analysis of Reference Samples by FI-ICP-MS. One of the most important ICP-MS features is its capability to perform very quick multielement analysis. However, when applying ICP-MS to the determination of hydride forming elements the analyst must be aware of the special chemistry required for the successful determination of hydride forming elements in more complex matrices. Depending on the sample type and method of preparation the accurate determination of all hydride forming elments in a single multielement ICP-MS run might not be possible.

For the study presented in this paper it was therefore decided to apply the optimized multielement program only to simple sample types such as water (NIST 1643b) and sea water (NASS-1). Unfortunately, both reference materials are not certified for all the hydride forming elements of interest (As, Se, Sb, Te, Hg and Bi). Therefore both samples were spiked with a few additional hydride forming elements at a concentration of 1ng/mL. The results of both analyses are summarized in Table 4. In the NIST 1643b water sample very good agreement between the measured and certified values was achieved. The recoveries for Sb and Hg (spike elements) were acceptable. Good results were also obtained in the analysis of the sea water sample.

Table 4: Results of the analysis of NIST 1643b water and NASS-1 sea water using hydride generation FI-ICP-MS; all concentration values in ng/L; * = spike

Element	Found (1643b)	Certified (1643b)	Range (1643b)	Found (NASS-1)	Certified (NASS-1)	Range (NASS-1)
As	52	(49)	--	1.85	1.65	0.19
Se	9.6	9.7	0.5	--	--	--
Sb	0.9	1.0*	--	0.93	1.0*	--
Hg	0.92	1.0*	--	0.9	1.0*	--
Bi	11.1	(11)	--	1.0	1.0*	--

Results of ETV-ICP-MS Single Ion Monitoring Determinations

Sensitivity and Precision. Sensitivity and precision experiments were performed for eight elements applying optimized single element ETV and spectrometer conditions. The elements were selected as being representative for the mass range typically covered by ICP-MS. The calculations were performed based on both, peak height and peak area measurements. Table 5 summarizes the results of these measurements.

Table 5: Results of single ion monitoring sensitivity and precision evaluation using an injection volume of 10µL and an analyte concentration of 100 ng/L; PkHt Sens. = mean Peak Height Sensitivity, PkHt % RSD = relative standard deviation in % for peak height measurements, PkArea % RSD = relative standard deviation in % for peak area measurements; 5 replicate ETV firings were made.

Element	Mn	Ni	Co	Cu	Rh	In	Pb	Bi
Pk Ht Sens.	6100	15500	7400	12600	1900	11200	110000	6000
Pk Ht %RSD	4.9	5.3	9	3.7	12	4.7	7.6	12
Pk Area %RSD	4.8	5.1	10	2.1	11	2.5	4.6	14

The results presented in Table 5 demonstrate that the measurement precision is only marginally improved comparing peak area measurements with peak height measurements. For elements which are relatively easy to atomize, such as Cu, In or Pb, better precision was achieved than for the more refractory elements, such as Co, Rh and Bi.

Detection Limits (Single Ion Monitoring). For the detection limit evaluation, 5 replicate measurements of 10 µL injections of 100 ng/L standards were performed, followed by 5 replicate measurements of 10 µL injections of a blank solution (1% HNO_3). From these measurements the typical 3-SD detection limits were calculated using the following equation (1).

$$D.L. = 3 * SD_{[Blank]} * Conc._{[Standard]} / Intensity_{[Standard]} \quad (1).$$

The results for all eight elements investigated are summarized in Table 6. All relative detection limits are lower than 7 ng/L (ppt). The best relative detection limit (120 fg/L) was achieved for Pb, which is not surprising since lead is one of the most sensitive elements in GF-AAS. In general the relative detection limits (10 µL injection volume) are only up to 5 times lower than those obtained with conventional solution analysis. Though, a further improvement in these relative detection limits is possible by increasing the sample injection volume from 10 µL to, e.g., 50 µL. Comparing the relative detection limits based on peak height measurements with those based on peak area measurements does not show any real differences. This means that the volatilization and transport efficiency is very reproducible from measurement to measurement. If this would not be the case, the peak height based detection limits should be poorer than those based on peak area measurements.

Table 6: ETV-ICP-MS detection limits (single ion monitoring); injection volume 10 µL, Pk Ht (rel.) = relative peak height detection limit, Pk Ht (abs.) = absolute peak height detection limit, Pk Area (rel.) = relative peak area detection limit, Pk Area (abs.) = absolute peak area detection limit

Element	Pk Ht (rel., ng/L)	Pk Ht (abs., fg)	Pk Area (rel., ng/L)	Pk Area (abs., fg)
Mn	2	20	2	20
Ni	5	50	7	70
Co	7	70	3	30
Cu	2	20	2	20
Rh	2	20	3	30
In	2	20	1	10
Pb	0.1	1	0.1	1
Bi	1	10	1	10

Results of ETV-ICP-MS Multielement Determinations

Sensitivity, Precision and Detection Limits. Sensitivity, precision and detection limits applying compromised multielement conditions were evaluated for 6 elements (Mn, Cu, Rh, In, Pb and Bi) and the results are summarized in Table 7. No real differences were found for peak area versus peak height data. As expected, the sensitivities are somewhat lower than those obtained in single ion monitoring. This is due to the fact that compromised ETV conditions were used which can not guarantee optimum volatilization conditions for each individual element.

For Cu and In only a minor difference in the absolute detection limits was found in comparison to the detection limits achieved under optimized single element conditions. For Mn, Pb and Bi the multielement detection limits differ from the single element detection limits by a factor of approximately 2. Only for Rh was a greater difference (factor 4) found. This is certainly due to the fact that the compromised ETV temperature/time program was far from being optimum for this element. The multielement detection limits are summarized in Table 7. The data demonstrate that multielement ETV-ICP-MS determinations for a group of elements (5 to 10) spread across the mass range can be performed without a dramatic loss in detection power. However, if many elements are determined in a multielement program the detection power must logically decrease. This is due to the fact, that the multichannel analysis must be performed with very short dwell times (< 1 ms) which has a considerable negative influence on the measurement precision at lower counting rates.

Table 7: ETV-ICP-MS multielement detection limits; injection volume 10 μL, Pk Ht (rel.) = relative peak height detection limit, Pk Ht (abs.) = absolute peak height detection limit

Element	Pk Ht (rel., ng/L)	Pk Ht (abs., fg)
Mn	4	40
Cu	2	20
Rh	6	60
In	2	20
Pb	0.3	3
Bi	2	10

ETV-ICP-MS Screening Analyses Using TotalQuant

ICP-MS spectra are relatively simple in comparison to, e.g., ICP-AES spectra. The information for nearly all elements of the periodic table are contained in only about 220 data points. These spectra are ideally suited to automated procedures of spectral interpretation. The ELAN TotalQuant software takes the information of previously scanned spectra and interprets their content in ways much like an experienced analyst might do. In this process, each isotope is evaluated for potential interferences, and TotalQuant´s internal heuristics dynamically guide the interpretation procedure so that the intensities assigned to each element are as accurate as they can be. Once intensities representing each of the 81 elements are computed, internal calibration information is used to estimate the element concentrations for all elements. The analyst usually updates these calibration data by running a blank and a simple multielement standard solution.

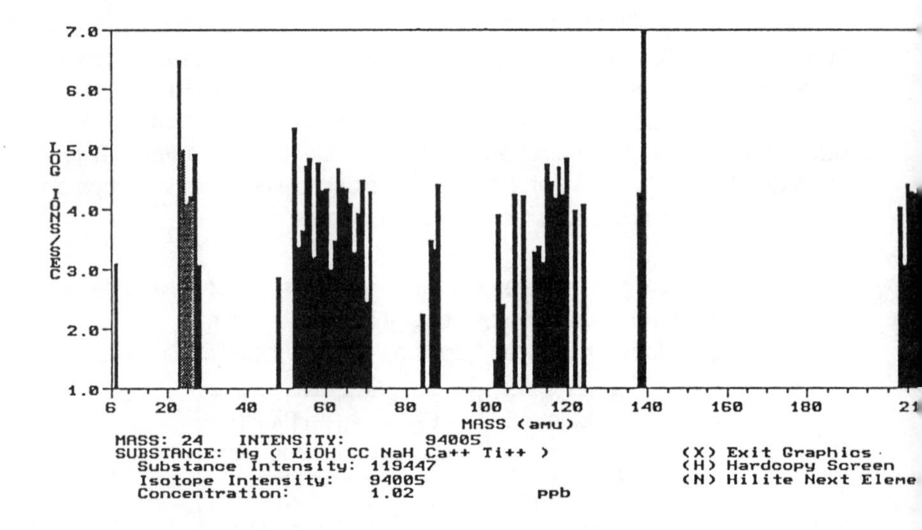

Figure 2: TotalQuant spectrum of a 1ng/L multielement solution from a single 10 µL ETV shot

For the ETV-ICP-MS TotalQuant screening analyses the instrument response table was updated using the data of a single ETV shot (10 µL injection volume) of a simple aqueous multielement solution. Figure 2 shows the spectrum of a 10 µL injection of a multielement solution (blank subtracted). For this application the mass range of the rare earth elements from 140 dalton to 180 dalton was skipped since these elements are not very efficiently vaporized in Massman-type graphite furnaces.

Figure 2 shows that ETV transient intensity data for many elements (full mass range) can be sampled with acceptable instrument sensitivity although non-optimum furnace condition must be applied for multielement ETV analyses. Table 8 summarizes the results of a further ETV-ICP-MS TotalQuant experiment in which both, the accuracy and reproducibility of the technique was investigated. Spectra of 5 replicate ETV shots (10µL injection volume of a 1µg/L multielement solution) were evaluated with the TotalQuant software. The mean recoveries and precision (SD) are presented in the table. In general, surprisingly good recoveries of the order of 80% to 110% were achieved. Only for Sr was the recovery somewhat poorer (60%).

Table 8: Mean recoveries and precision data (SD) for 5 replicate ETV shots of a 1µg/L multielement standard solution (injection volume 10µL)

Element	Recovery (%)	SD	Element	Recovery (%)	SD
Li	109	25	Ga	104	21
Na	100	10	Sr	60	8
Mg	91	21	Rh	94	20
Al	96	10	Ag	109	21
Mn	89	14	Cd	110	16
Co	82	6	In	119	23
Ni	84	8	Tl	96	19
Cu	98	23	Pb	106	18
Zn	100	18	Bi	92	20

Finally a certified reference sample (NRC SLRS-1) was analyzed. The results are summarized in Table 9. With the exception of Fe, the recoveries were in the order of 60% to 140%. The much higher recovery for Fe of 250% was probably due to a contamination problem.

3 CONCLUSION

The application of alternate sample introduction techniques for ICP-MS offers a number of significant analytical advantages over conventional sample introduction by solution nebulization. Flow injection is a very promising technique which allows to perform online matrix separation as well as analyte preconcentration followed by multielement ICP-MS analysis. The results for hydride generation FI-ICP-MS presented in this study demonstrate that flow injection can also be applied very successfully to improve the detection power of ICP-MS for elements such as As, Se and Hg

Table 9: Mean recoveries and precision data (SD) for 5 replicate ETV shots of Riverine Water SLRS-1 (injection volume 10μL)

Element	Recovery (%)	SD	Element	Recovery (%)	SD
Al	108	26	Cu	61	7
Mn	132	23	Zn	66	7
Fe	250	35	Cd	150	37
Ni	59	8	Pb	73	25

for which conventional ICP-MS (solution aspiration) is sometimes not sensitive enough. Furthermore, due to the matrix-analyte separation involved in hydride generation flow injection, As, Se and other hydride forming elements can be determined directly in samples with high concentrations of dissolved solids such as sea water.

Electrothermal vaporization is another promising alternative sample introduction technique for ICP-MS for either quick overview analysis or special applications. If only a few elements are to be determined simultaneously, ETV-ICP-MS allows the determination of elements at femtogram levels. Due to the volatilization of the water matrix prior to the vaporization of the analyte elements of interest, a number of metal oxide interferences can be eliminated. The technique can further be used for the analysis of solid samples applying either the cup-in-tube technique (powders and small particles) or the slurry sampling technique (powders). Some conventional graphite furnace atomizers for AAS can be easily converted into electrothermal vaporizers by means of a simple modification kit. The advantage of using a modified conventional heated graphite furnace atomizer (HGA) is that all accessories which are commercially available for these furnaces (autosamplers, high quality pyrolytically coated graphite tubes, solid sampling accessories, etc) can be used.

References

[1] J.J. Thompson, and R.S. Houk,, Anal. Chem., 1986, 58, 2541

[2] H.M. Crews, J.R. Dean, L. Ebdon, R.C. Massey, Analyst, 1989, 114, 895

[3] H. Klinkenberg, S. van der Wal, J. Frusch, L. Terwint and T. Beeren, At. Spectrosc., 1990, 5, 198

[4] P. Arrowsmith, Anal. Chem., 1987, 59 (10), 1437

[5] J.W. Hager, Anal. Chem., 1989, 61(11), 1243

[6] C.W. McLeod, ICP Inf. Newsl., 1987, 12, 10, 721

[7] J.R. Dean, L. Ebdon, H.M. Crews, R.C. Massey, J. Anal. At. Spectrom., 1988, 3, 349

[8] R.C. Hutton, A.N. Eaton, J. Anal. At. Spectrom., 1988, 3, 547

[9] X. Wang, M. Viczian, A. Lasztity and R.M. Barnes, J. Anal. At. Spectrom., 1988, 3, 821

[10] M. Janghorbani, and B.T.G. Ting, Anal. Chem., 1989, 61, 701

[11] Y. Israel, A. Lasztity, and R.M. Barnes, Analyst, 1989, 114, 1259

[12] C.J. Park, and G.E.M. Hall, Geol. Surv. Can. Pap., 1986, 86, 767

[13] C.J. Park, and G.E.M. Hall, J. Anal. At. Spectrom., 1987, 2, 473

[14] C.J. Park, J.C. Van Loon, P. Arrowsmith, and J.B. French, Anal. Chem., 1987, 59, 2191

[15] C.J. Park, J.C. Van Loon, P. Arrowsmith, and J.B. French, Can. J. Spectrosc., 1987, 32, 29

[16] A.R. Date, and Y.Y. Cheung, Analyst, 1987, 112, 1531

[17] U. Voellkopf, paper presented at the 1988 Winter Conference on Plasma Spectrochemistry, San Diego, Cal, USA, 1988

[18] C.J. Park, and G.E.M. Hall, J. Anal. At. Spectrom., 1988, 3, 355

[19] G.E.M. Hall, C.-J. Pelchat, D.W. Boomer, and M.J. Powel, J. Anal. At. Spectrom., 1988, 3, 791

[20] D.C. Gregoire, J. Anal. At. Spectrom., 1988, 3, 309

[21] W.-L. Shen, J.A. Caruso, F.L. Fricke, R.D. Satzger, J. Anal. At. Spectrom., 1990, **5**, 451

[22] J. Ruzicka, and E.H. Hansen, Anal. Chim. Acta, 1975, 78, 145

[23] "FIAstar, Flow Injection Analysis, Bibliography", Tecator, Höganas, Sweden

[24] D.E. Nixon, V.A. Fassel, and R.N. Knisley, Anal. Chem., 1974, 46, 210

[25] G. Crabi, P. Cavalli, M. Achilli, G. Rossi, and N. Omenetto, At. Spectrosc., 1982, 3,4, 81

[26] U. Voellkopf, Z. Grobenski, R. Tamm and B. Welz, Analyst, 1885, 110, 573

[27] U. Voellkopf, R. Lehmann and D. Weber, Analyst, 1987, 2, 455

A Continuous Hydride-Generation System for ICP-MS Using Separately Nebulized Internal Standard Solutions

Paul G. Ek[1], Stig G. Huldén[1], Erland Johansson[2], and Torsten Liljefors[2]

[1] LABORATORY OF ANALYTICAL CHEMISTRY, ÅBO AKADEMI UNIVERSITY, 20500 ÅBO, FINLAND
[2] DEPARTMENT OF RADIATION SCIENCES, DIVISION OF PHYSICAL BIOLOGY, UNIVERSITY OF UPPSALA, BOX 535, S-751 21 UPPSALA, SWEDEN

During the last decade inductively coupled plasma mass spectrometry (ICP-MS) has shown a remarkable delvelopment as a new powerful tool for performing elemental and isotopic analysis[1-8]. Its high power of detection, broad linear dynamic working range, multi-element capability, simplicity of spectrum and possibilty of performing isotopic analysis makes it an attractive addition to the established spectrometric techniques (i.e. AAS, ICP-AES, XRF) for many applications.

In spite of the versatility of the ICP-MS technique there are in fact some important applications where inadequate sensitivity due to complex background spectra and infavorable isotopic conditions restrict the analytical performance of the technique. For example the accurate determination of selenium in biological and environmental samples is of great importance due to the obvious role of selenium for human and animal health[9,10].

Because of the low selenium concentrations (ng g^{-1} - μg g^{-1}) normally present in biological materials a very high sensitivity is required of the analytical method used. However, although ICP-MS shows detection limits in the 0.01 - 0.1 μg l^{-1} range for most elements the detection limit for selenium is not better than 0.8 μg l^{-1}, when solution nebulisation is employed[5]. This is related to factors such as the relatively high ionisation potential (9.75 eV) of selenium and isotopic interferences, the major selenium isotope (^{80}Se) being masked by the argon ($^{40}Ar_2{}^+$) dimer. Polyatomic ions originating from matrix elements present at higher concentrations than the analyte element may cause serious spectral interferences in determination of selenium in complex matrices[11]. Also the prescence of easily ionized elements may severely suppress the analyte signal[4,5].

In order to reduce these physicochemical effects several strategies, either alone or in combination, are available: (1) the method of standard addition, (2) the isotope dilution technique, (3) chemical isolation of the interferents, (4) the use of internal standard and (5) chemical isolation of the analyte. The genaration of volatile hydrides of elements such as arsenic, bismuth, selenium etc. in order to separate and preconcentrate the analyte from the sample matrix is a well established technique for improving sensitivity in atomic spectroscopy[12,13]. This alternative sample introduction technique was first adopted for ICP-MS by Date and Gray[14]. Later Powell et al.[15] showed that by continuous hydride generation greatly improved detection limits could be obtained for several elements compared with pneumatic nebulization. Wang et al.[16] have investigated the technique for determination of lead by isotopic dilution. Janghorbani and Ting[17] and Ting et al.[18] demonstrated the applicability of continuous hydride generation for isotopic determination of selenium in biological materials using isotope dilution for quantification.

The aim of this study was to develope a rapid and accurate method for routine determination of volatile hydride forming elements at trace levels. This paper describes a continuous hydride generation system for ICP-MS analysis and its application for determination of selenium. Although isotope dilution is a standard procedure in mass spectrometry and the most accurate method for performing quantitative measurements, it is more tedious than other calibration strategies, e.g. external calibration. Its use is restricted to elements that have at least two stable natural isotopes and also by the availability of specially enriched isotope standards of the analyte elements.

In this work we therefore have investigated the use of external calibration for HG-ICP-MS determination of selenium. To compensate for potential changes in instrumental sensitivity of the ICP-MS ^{115}In was used as internal standard. Owing to the poor hydride forming properties of indium[19] we have focused on the possibility of continuous introduction of the indium standard solution separately into the plasma by pneumatic nebulisation via the spray chamber. The use of a properly hydride forming element (Te) as internal standard has also been tested.

The reliability of the proposed method was assessed by determination of the selenium content of several biological materials including: NBS SRM 1577a (Bovine Liver), SRM 1567a (Wheat Flour), SRM 1566 (Oyster Tissue) and IAEA (International Atomic Energy Agency) H-4 (Animal Muscle) and H-5 (Animal Bone).

Experimental

The Hydride ICP-MS analysis system

The instrument employed in these studies was a commercially available ICP-MS system (Plasma Quad, PQ 1, VG Isotopes, Winsford, Cheshire, UK). The major instrument components are schematically shown in Fig. 1a. The interface between the plasma and the mass spectrometer consists of nickel sampling and skimmer cones with orifice diameters of 1.0 and 0.75 mm, respectively. The nebuliser used was a Meinhard C Concentric and the plasma torch was of a Fassel design. Sample uptake rate was controlled by means of a multichannel peristaltic pump. For data acquisition the Plasma Quad software (3.1A) supplied by the instrument manufacturer was utilized.

For hydride determination a laboratory-built continuously operating hydride generator was used (Fig. 1b). The construction and operation of the continuous hydride-generator, originally developed for a DCP-AES, has been described elsewhere[20]. For Hydride-ICP-MS operation this hydride generation system was scaled down about 40 % in order to reduce sample and reagent consumption.

The hydride generator module can be easely attached to the ICP-MS spectrometer by connecting the hydride introduction tube (5 mm i.d. PTFE tube, length 1.2 m) to a separate T-junction mounted on the injector channel of the ICP - torch. The purpose of the relatively long reaction coil (length 1 m, i.d. 2.7 mm) is to reduce the effect of the small pressure fluctuations generated by the peristaltic pump, which are further amplified by the evoluted hydrogen gas from the $NaBH_4$ - HCl reaction. A smooth transfer of the gaseous reaction products to the plasma is of outmost importance to get a stable response signal.

Reagents

The reagents used were prepared from analytical reagent grade chemicals without any further purification. All solutions were made up in demineralized water obtained from an ELGA water purification system.

The reductant solution, 1% sodium borohydride solution was prepared by dissolving 10 g of $NaBH_4$ pellets (Spectrosol R, BDH Chemicals) and 4 g of sodium hydroxide (Merck p.a.) in 100 ml of water. In order to remove the resulting carbonate precipitate the solution was filtered through a 0.45 μm

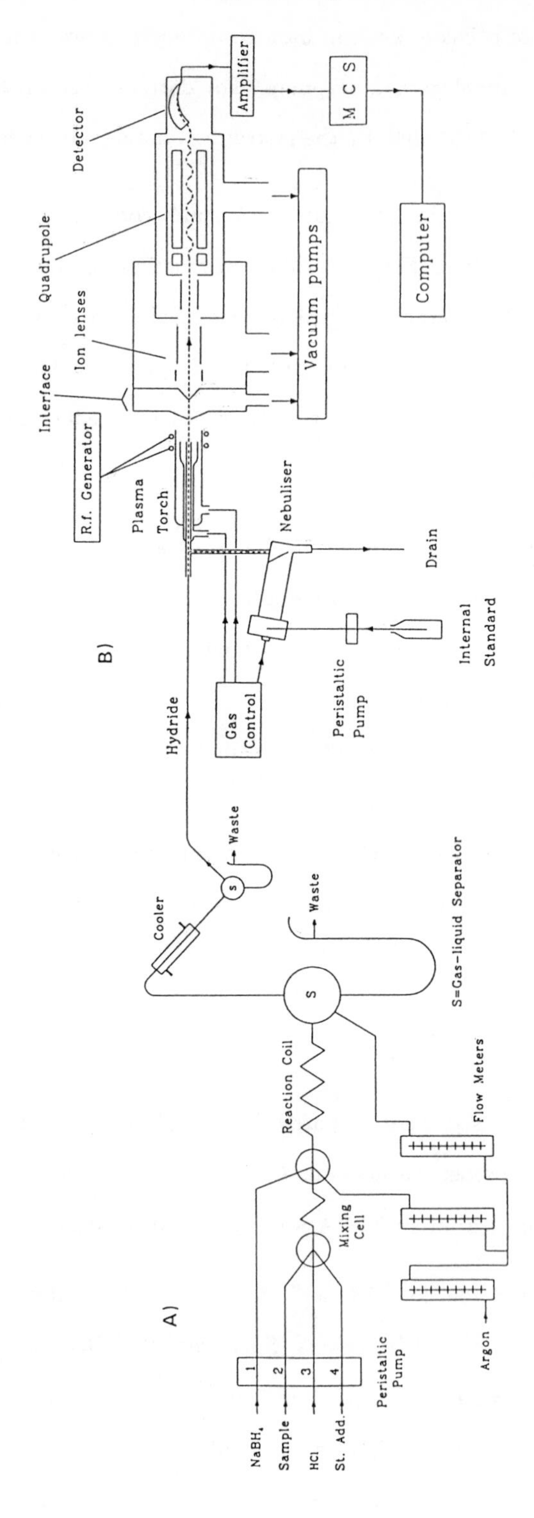

Fig. 1. Schematic diagram of the Hydride Generation - Inductively Coupled Plasma Mass Spectrometer System.

membrane filter. The clear solution was diluted to 1000 ml and stored in a polyethylene bottle in the dark. A 5 M hydrochloric acid solution prepared from Merck:s p.a. 37 % HCl, was used for continuous acidification of all test solutions according to the flow chart given in Fig.1b.

The internal standard solutions (100 $\mu g\ l^{-1}$) of indium and tellurium were prepared by dilution from 1.00 mg ml^{-1} stock solutions (Spectrosol, BDH Chemicals Ltd) of indium nitrate and telluric acid, respectively. The indium solution was made 5 % (v/v) in nitric acid and the tellurium solution 1 % in hydrochloric acid.

The standard working solutions of Se(IV) were made up by appropriate dilution with 0.1 M hydrochloric acid of a 1.0 mg/ml Se(IV) stock solution, (Johnson and Matthey, Specpure).

A 0.1 M solution of 1.10-phenanthroline hydrochloride (Merck p.a.) was prepared as masking agent for copper. Suprapur 65 % nitric acid (Merck) and 70 % perchloric acid were used for dissolution of the samples. For reduction of Se(VI) to Se(IV) the samples were heated to boiling in 6 M hydrochloric acid.

Sample preparation

In most biological materials selenium is incorporated as very acid-resistant organoselenium compounds[21], which require extreme conditions for their decomposition. The samples were therefore decomposed by a wet ashing procedure slightly modified from the method described by Franck[22], using an automatic Tecator 12 (Tecator AB, Höganäs, Sweden) digestion system. About 200 - 300 mg of the samples were accurately weighed into the digestion tubes (glass tubes with 24 mm bore). After addition of 10 ml HNO_3/ $HClO_4$ mixture (7+3 v/v) the samples were digested for 12 h (overnight) by a stepwise increase of the ashing temperature to 215 °C. After cooling, 10 ml of HCl was added and the solutions were heated to gentle boiling for 10 min

Table 1. Operating conditions of HG-ICP-MS system.

IC-Plasma -	
r. f. forward power	1400 W
Reflected power	< 10 W
Coolant gas flow rate	13 l min^{-1}
Auxiliary gas flow rate	0.5 l min^{-1}
Nebuliser gas flow rate	0.70 l min^{-1}
Flow rate of internal st.	1 ml min^{-1}
Spray chamber temp.	10 °C
Hydride Generator -	
Sample flow rate	6.6 ml min^{-1}
HCl (5M) flow rate	1.2 ml min^{-1}
$NaBH_4$ (1%) flow rate	1.2 ml min^{-1}
Carrier gas flow rate	90 ml min^{-1}
Peak jump conditions -	
Isotopes (selected)	^{78}Se, ^{82}Se, ^{115}In
Number of channels	1024
Dwell time	^{78}Se 327680 μs
	^{82}Se 327680 μs
	^{115}In 10240 μs
Point per peak	5
Peak jump sweeps	20
Collector type	Pulse

to convert selenate to selenite. The NBS samples were transferred to 100 ml standard flasks and diluted to volume with water. Because of the relatively low selenium contents in the IAEA reference samples they were diluted to 25 ml.

The NBS SRM 1643b (Trace elements in water) was diluted 4-fold prior to analysis with water to reduce the nitric acid concentration (5 % v/v) of the sample. Nitric acid and especially its reduced derivatives (NO_2^-, NO_x) are serious inhibitors of H_2Se evolution[23].

HG-ICP-MS analysis

The ICP-MS spectrometer is changed from normal to hydride operation mode simply by connecting the hydride outlet tube to the detachable T-junction of the plasma injector channel as previously described.

After the normal starting up procedure the ICP-MS is tuned for maximum sensitivity on mass ^{115}In while nebulising a 100 μg l^{-1} standard solution. Before the peristaltic pump of the hydride generator is started the carrier gas valves are opened and the argon and the liquid flow rates are set to the recommended values (Table 1).

Results and Discussion

Investigation of operating conditions

The effect of interfacing the hydride generation accessory to the ICP mass spectrometer was expected to have only a small influence on its normal operation. The most striking effect observed was a slight increase of the reflected power, from the normal value of < 5 W to 8 - 10 W. However, in order to maximize the analytical sensitivity and precision parameters known to be most critical, i.e. parameters associated with plasma operation and sample introduction, have to be examined. By a univariate search we investigated the effect of varying such parameters as the nebuliser- and auxilary gas flow rates, the power coupled to the plasma, the sample solution uptake rate, the hydride carrier gas flow rate and the concentration and flow rates of the sodium-borohydride and hydrochloric acid. Figures 2 - 6 illustrate some of parameters studied and their effect on the analyte signal.

Carrier gas flow rate

One of the most critical parameters studied was the carrier gas flow rate, (Fig. 2.) because it has a direct influence on the selenium signal intensity. At higher carrier gas flow the hydride is more effectively expelled from the solution.

Although an optimum signal to carrier gas flow rate ratio was found at 0.12 l min^{-1}, the gas flow rate had to be lowered for practical work. This because

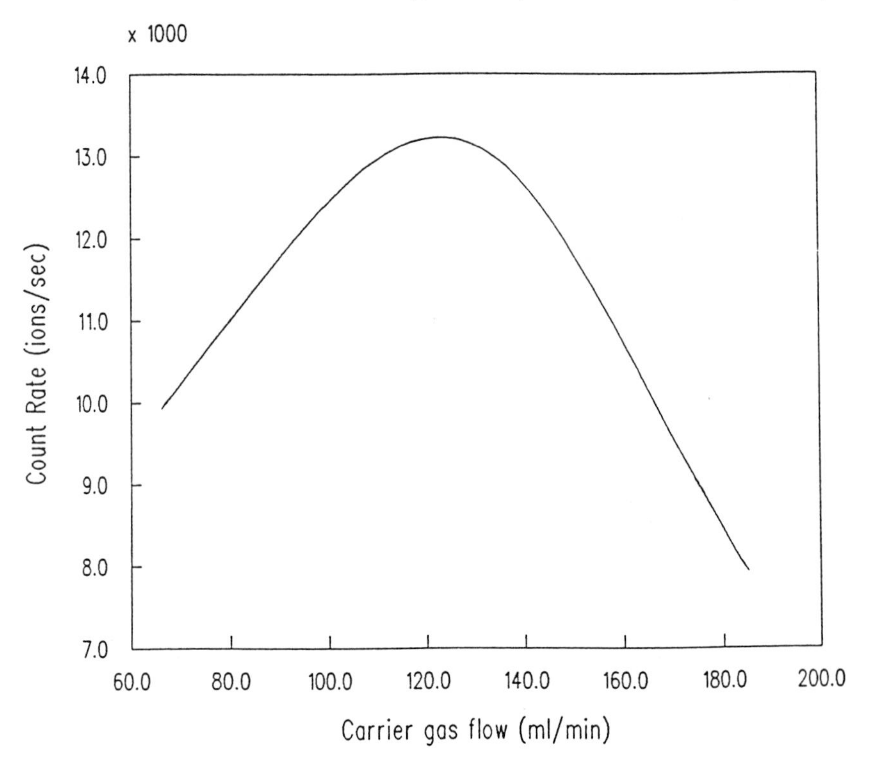

Fig. 2. Effect of carriergas flow rate on Se ion intensity. (5 μg l^{-1} Se, nebuliser flow 0.7 l min^{-1}, forward power 1400 W. For other parameters see Table 1.)

the response of internal standard drastically dropped due to the dilution effects (Fig. 3.). Raising the gas flow to 0.19 l min^{-1} further decreased the ^{115}In signal to about 14 kcps (kilocounts per second). When starting the pump again the signal increased to about 50 kcps. This change of the observed signal is probably due to an increase of the plasma temperature caused by the evolved hydrogen gas.

As a compromise the carrier gas flow rate was set at 0.09 l min^{-1} to keep the ^{115}In ion intensity at a reasonably high level of 60 - 70 kcounts s^{-1}.

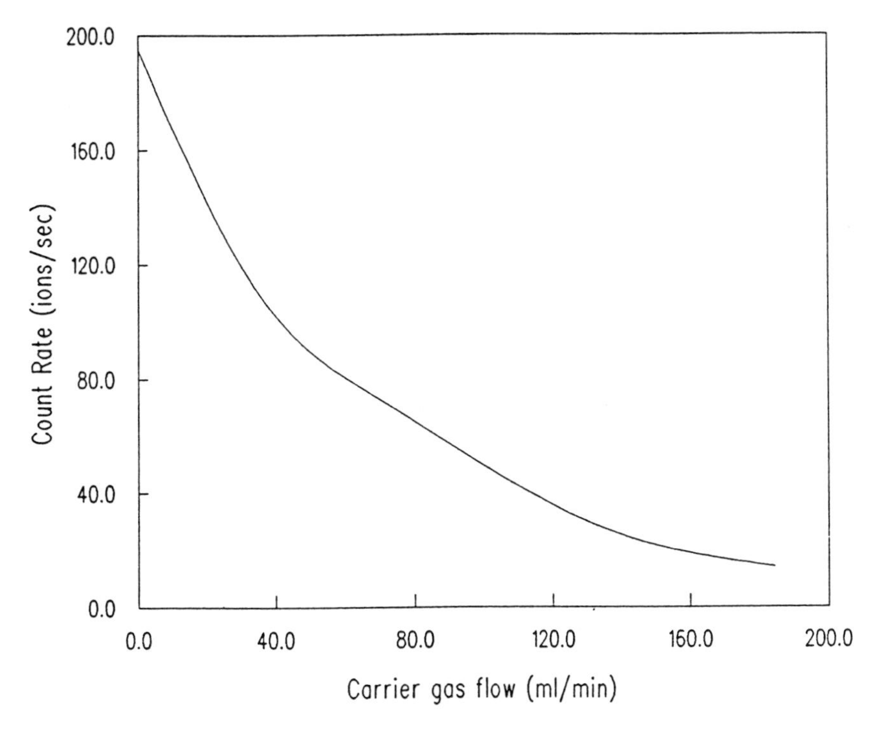

Fig. 3. Effect of carrier gas flow rate from the hydride generator on internal standard (100 μg l^{-1} In) ion intensity, when the hydride generator pump is not in function. (Nebuliser gas flow 0.7 l min^{-1}, forward power 1400 W. Other parameters according to Table 1.)

Sample uptake rate

The sample solution uptake rate versus analyte signal response is illustrated in Fig. 4. There is an almost linear relationship between the sample introduction rate and the analyte response from 6.5 up to 11.5 ml min^{-1}. The flow rate in the range 0 - 6.5 ml min^{-1} was not examined because the time needed to transfer the sample and the reagent solutions to the first reaction cell turned out to become too long for practical work. On the other hand the sensitivity is reduced at very low sample flow rates. By changing to thinner pumping tubes, however, the speed of the liquid flow is increased and smaller sample volumes can thus be analysed.

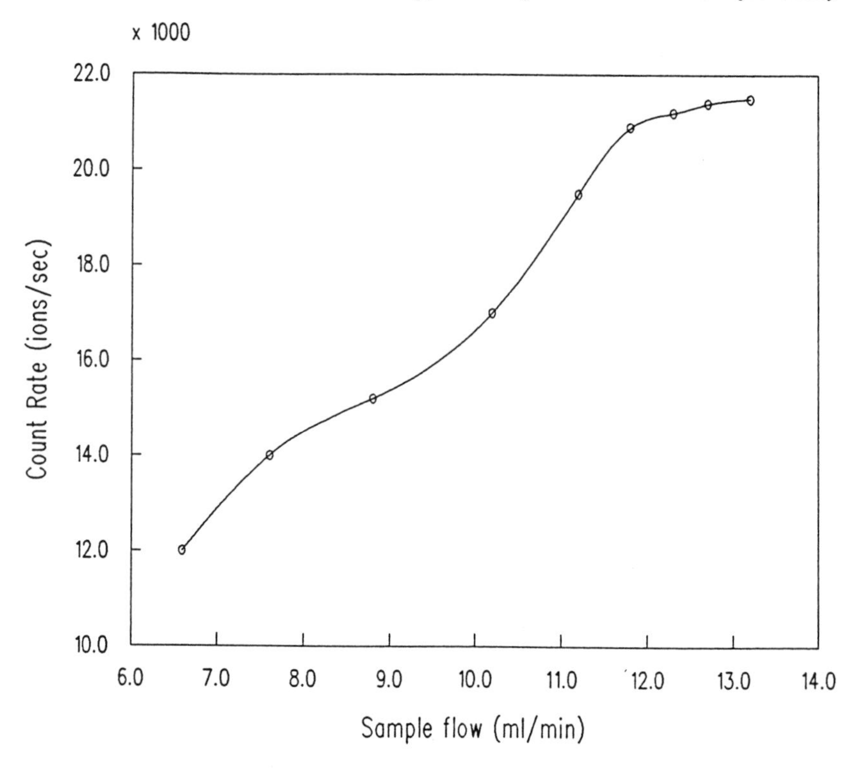

Fig. 4. The influence of sample uptake rate on analyte ion intensity. (5 μg l^{-1} Se, For other parameters see Table 1).

In order to reduce sample consumption, most of the experiments in this study were made with a sample and reagent flow rate of 6.6 and 1.2 ml min^{-1} respectively. With this sample flow rate and a carrier gas flow rate of 0.09 l min^{-1}, a stable response signal was obtained within 20 s of introduction of the sample solution. The flexible sample flow rate is one of the main advantages of continuous hydride generation. Because of the constant ratio between the sample flow rate and the reagent flow rate and if the available sample volume is not a limiting factor, lower or higher analyte concentration can be determined by inceasing or decreasing the pumping rate without affecting the solution conditions.

Forward power

Fig. 5. shows the effect of the incident power on the analyte response. The count rate increases with the applied power to the plasma. This pattern is similar at all the measured concentration levels. No clear maximum of the intensity of the ^{82}Se-signal could be observed within the measured power range.

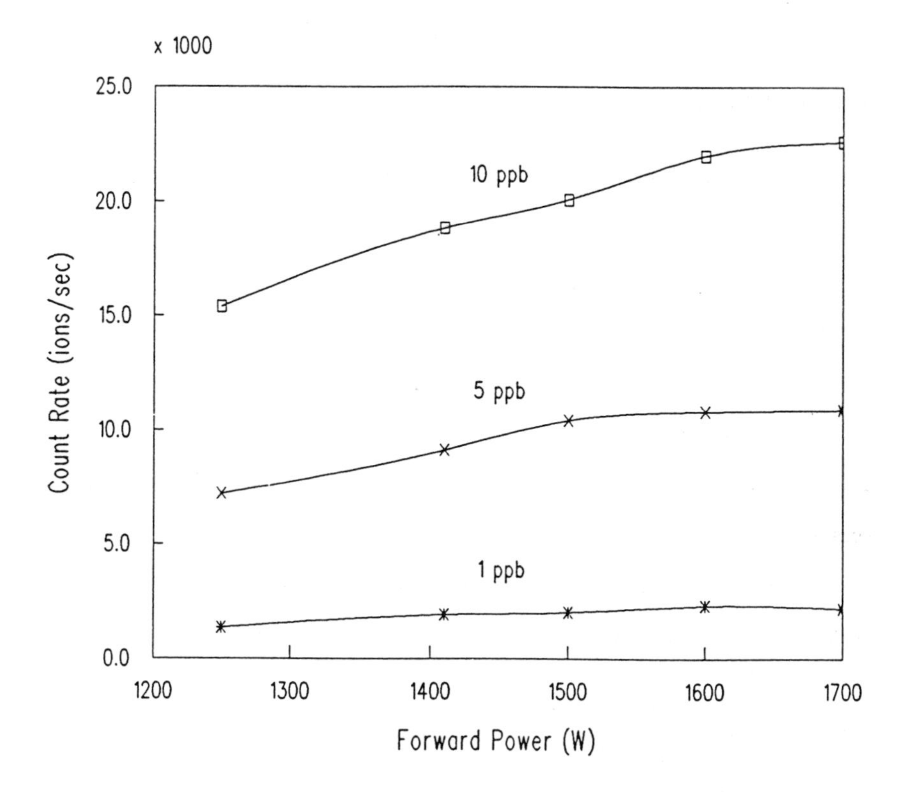

Fig. 5. Effect of forward power on analyte ion intensity measured at three different concentration levels. (For parameters see Table 1).

Since it was found that higher power levels also increased the background signal, a forward power of 1400 W was selected for the measurements.

Sodium Borohydride and Hydrochloric acid concentration.

The effect of the reductant concentration and the acid molarity on the analyte response (5 μg l^{-1} Se) was studied in the range 0.1 - 3 % and 0.5 - 6 M, respectively. Maximum analyte signal/background ratio was obtained at a $NaBH_4$ concentration of 1 % (w/v) and at a HCl concentration of 3 M. The sample and reagent flows were kept constant at 6.6 and 1.2 ml min^{-1}. The results were identical with those obtained previously[20].

Effect of Internal Standard

The utilization of internal standardisation is a necessity for accurate ICP-MS work. It is used to correct for signal instability caused by e.g. fluctuations in the sample matrix, potential changes in the temperature of the plasma, changes in aerosol transport, nebuliser and sampler orifice blockages etc. which all result in instrumental drift. In HG-ICP-MS analysis the use of internal standards is especially important because even small fluctuations in the gas transport (hydride + hydrogen + carrier gas) may affect the plasma temperature and hence the ion population. ^{115}In is an accepted internal standard in ICP-MS work. In the initial experiments carried out, blanks, samples and standards were spiked with 100 μg l^{-1} indium standard solution as an internal standard for selenium determination. Due to the poor hydride forming properties[19] the signal response was low, in the order of 5000 counts s^{-1} and the experiment was therefore abandoned. By spiking the solutions with 100 μg l^{-1} tellurium (^{126}Te), a good hydride forming element, a response of 90 kcps was obtained. In determination of selenium at 10 μg $^{-1}$ and 100 μg l^{-1} levels on masses ^{77}Se, ^{78}Se and ^{82}Se the following results were obtained.

Sample	^{77}Se	^{78}Se	^{82}Se
10 μg l^{-1} Se	11.34 ± 0.69	11.87 ± 0.87	11.34 ± 1.03
100 μg l^{-1} Se	121.79 ± 2.97	120.60 ± 2.73	121.37 ± 3.34

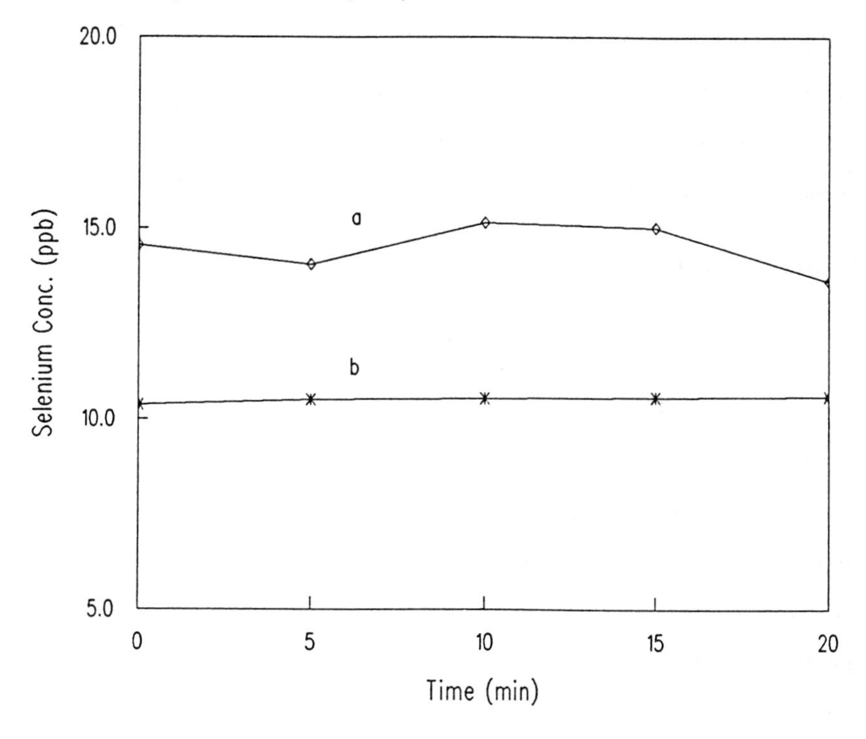

Fig. 6. Effect of internal standard on the analyte ion (10 $\mu g\ l^{-1}$ Se) signal stability vs. analysis time. Internal standard introduced externally via spray chamber.

A. Introduction of 1 ml min^{-1} water instead of internal standard.

B. Introduction of 1 ml min^{-1} 100 $\mu g\ l^{-1}$ In standard solution.

Operating parameters according to Table 1.

The results indicate a high positive error, but the reason for this was not further investigated.

By addition of the internal standard externally via the spray chamber also non-hydride forming elements can be used as internal standards for HG-ICP-MS. Fig. 6 b and Table 2 b show the results obtained for determination of 10 $\mu g\ l^{-1}$ Se using ^{115}In as a separately introduced internal standard. The selenium concentration was measured every fifth minute during the experiment.

From Fig. 6 it can be seen that the long-term (20 min) stability is good. The drift was calculated to +0.8 %. On the other hand a very high positive error was obtained in the test without internal standard present and the response signal was instable. The separate addition of ^{115}In as internal standard was therefore applied in the subsequent measurements.

Analytical performance

Linearity and detection limit

To assess the performance of this method laboratory standard solutions of Se(IV) were analysed under optimum conditions (Table 1). Calibration graphs were prepared by measuring selenium concentrations of 0, 1, 5 and 10 μg l^{-1} in the peak jumping mode on masses ^{78}Se and ^{82}Se. In Table 2 the corresponding calibration data, measured without and with the internal standard, are given. Table 2 also includes the results for some selenium standard solutions of the same run. As can be seen from the calibration data the linearity of the calibration curve is bad when no internal standard was used, while a very excellent linearity in the measured concentration range was obtained in the second case. The slope of the calibration curve for ^{78}Se was 4430 and that for ^{82}Se 1760 cps μg^{-1} l^{-1}, respectively.

The detection limit, calculated as three times the standard deviation of the blank was 0.05 μg l^{-1} for both isotopes. By reducing the reagent blank values ($NaBH_4$), the detection limit may further be lowered. The precision (r.s.d.) calculated from several individual runs varied between 4 - 12 %, 0.8 - 2.8 % and 0.6 - 1.6 % at 0.25, 2.5 and 10 μg l^{-1} concentration levels.

Interferences

Due to the chemical separation of the analyte from most of the concomitant matrix elements spectral interferences (isobaric overlapping) and non-spectroscopic interferences (signal suppression) are reduced in the HG-ICP-

Table 2. The effect of externally added internal standard in HG-ICP-MS analysis. A comparison of calibration data and analytical results obtained without and with internal standard present. (Calibration data obtained by analysis of 0, 1, 5 and 10 μg l^{-1} Se standard solution).

Calibration curve equation: $y = a + bx + cx^2$

A. Peak Jump Analysis of Selenium without Internal Standard.

Calibration parameters:

Element	Mass	a	b	c	r
Se	78	-1064.16	3795848.70	0.00	0.98866
Se	82	-408.29	1509028.36	0.00	0.98847

Analysis results (n=3):

Samples	Expected μgl^{-1}	Measured on ^{78}Se μgl^{-1}	Measured on ^{82}Se μgl^{-1}
Stand. 1	0.25	0.64 ± 0.03	0.69 ± 0.02
Stand. 2	2.5	3.4 ± 0.5	3.5 ± 0.4
Stand. 3	10.0	13.9 ± 1.3	14.1 ± 1.3

B. Peak Jump Analysis of Selenium with Internal Standard.

Calibration parameters:

Element	Mass	a	b	c	r
Se	78	-28.60	4435889.25	0.00	1.00000
Se	82	-36.62	1761940.86	0.00	0.99998

Analysis results (n=3):

Sample	Expected μgl^{-1}	Measured on Se 78 μgl^{-1}	Measured on Se 82 μgl^{-1}
Stand. 1	0.25	0.26 ± 0.03	0.25 ± 0.03
Stand. 2	2.5	2.52 ± 0.03	2.49 ± 0.02
Stand. 3	10.0	10.35 ± 0.06	10.42 ± 0.16

MS system. Similarly the background spectra generated by polyatomic ions derived from solvent acids are diminished.

On the other hand chemical interferences in the hydride generation step is a drawback of the HG-ICP-MS technique. The presense of certain transition metal ions of groups VIII and IB (e.g. Ni(II), Cu(II), Ag(I) etc.) severely

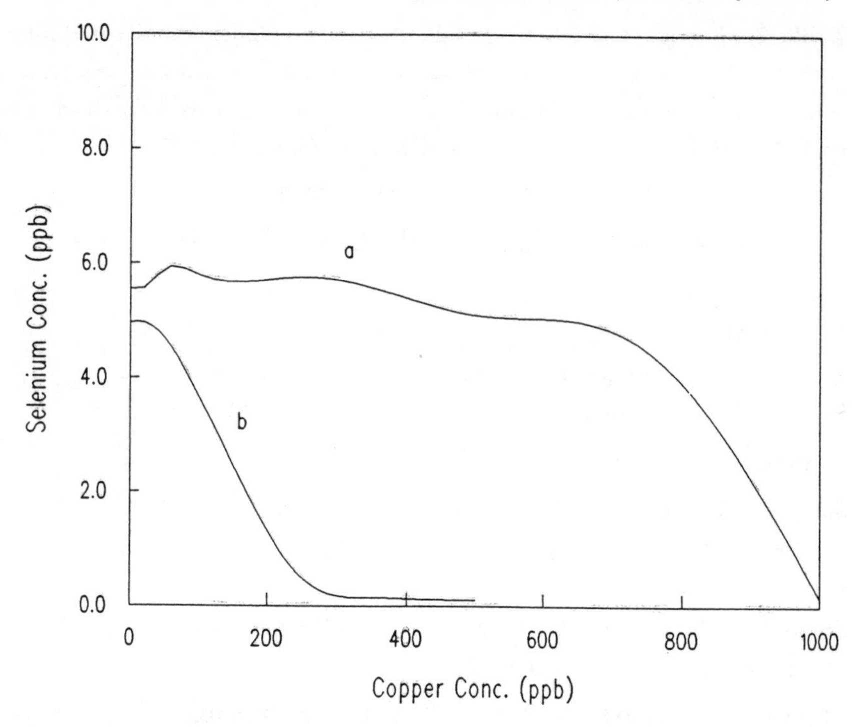

Fig. 7. The effect of copper(II) on selenium hydride signal ($5 \mu g\ l^{-1}$ Se).

A. Effect of on-line addition of 0.1 M 1,10-phenanthroline ($1.2\ ml\ min^{-1}$) as masking agent for copper.

B. Influence of copper(II) concentration on the recovery of selenium. For analytical conditions see Table 1.

inhibit the evolution of the hydrides[24]. In the present work we have mainly studied the interference of copper in the determination of selenium. Fig. 7b describes the effect of different concentrations of copper on $5\ \mu g\ l^{-1}$ selenium. The threshold concentration at which Cu(II) starts to interfere is about 0.1 $mg\ l^{-1}$. By using masking agents such as 1,10-phenanthroline the tolerance limit for copper can be increased to about 0.6 $mg\ l^{-1}$ Fig. 7a. These results are in concordance with those obtained in a previous study[20].

Another source of interference affecting the hydride generation step are oxidizing acids such as nitric acid which may be present as solvent rests from the sample dissolution procedure. In the presence of copper the inhibition effect of HNO_3 and its reduced derivatives (NO_2^-) is further strengthened[23].

Analysis of reference materials

The performance and the reliability of the HG-ICP-MS method was evaluated by analysis of the certified biological reference materials. In the initial experiments very low recoveries were obtained. For instance for the NBS 1566 sample the selenium recovery was only 52 % at a copper concentration of 270 μg l^{-1}. Attempts to mask the effect of copper by on-line addition of 1,10-phenontraline only enhanced recovery to 70 %.

Table 3. Determination of selenium by HG-ICPMS using separate nebulisation of the internal standard (100 ppb In).

	Se concentration /μg l^{-1}		
Sample	Expected	Measured	Recovery %
0.25 μgl^{-1}	0.25	0.23 ± 0.01*	92
1.0 μgl^{-1}	1.0	0.99 ± 0.03*	99
2.5 μgl^{-1}	2.5	2.46 ± 0.07*	98
10.0 μgl^{-1}	10.0	10.37 ± 0.11†	104
NBS 1643b°	9.7 ± 0.5	9.1 ± 0.3†	94
	Se concentration /μg g^{-1}		
NBS 1577a	0.71 ± 0.07	0.90 ± 0.05†	127
NBS 1567a	1.1 ± 0.2	1.04 ± 0.02 †	95
NBS 1566	2.1 ± 0.05	2.03 ± 0.04†	97
IAEA H-4	0.28 ± 0.03	0.27 ± 0.02†	96
IAEA H-5	0.054 ± 0.013	0.056 ± 0.017†	105

* mean of five determinations.
† mean of three determinations.
° diluted 4-fold before analysis.

Another strategy was therefore tested based on the excellent sensitivity of the HG-ICP-MS technique. By diluting the NBS samples (1577a, 1567a and 1566) with water (1+9) the concentration of copper and HNO_3 is reduced to such a low level were no interference is possible. As already described the NBS 1643 b sample was diluted (1+3) in order to decrease the nitric acid concentration. The results are presented in Table 3 showing a fairly good agreement with the certified (NBS samples) and the recommended (IAEA samples) values.

Owing to the very low selenium concentration in the IAEA reference samples they were diluted to 25 ml in the sample preparation step. The good recovery obtained at this high a solute concentration was possible due to the low copper content in these materials.

Conclusions

The combination of continuous hydride generation with ICP-MS offers a sensitive method for determination of selenium. By using external calibration and a separate introduction of internal standard, rapid and reproducible determinations of selenium at trace levels can be performed. Due to the chemical separation of the analyte from the matrix many spectral interferences typical for normal solution nebulisation are reduced. Reduction at the drawbacks of the hydride generation step including transition metal interference on hydride evolution, is obtained by using a masking agent or simply by dilution of the sample when possible. The results obtained for determination of selenium in biological standard reference materials were satisfactory indicating the potential of the method for other types of real samples.

The authors acknowledge with gratitude the donation of Crafoord foundation making this study possible. P.G.E. and S.G.H. acknowledge a grant from the Åbo Akademi University which enabled their research at the University of

Uppsala. E.J. is indebted to Astra Läkemedel AB, Södertälje for financial support.

References

1. Houk, R. S., Fassel, V. A., Flesch, G. D., Svec, H. J., Gray, A.L. and Taylor, C. E., *Anal. Chem.*, 1980, **52**, 2283.
2. Houk, R. S., and Thompson, J. J., *American Mineralogist*, 1982, **67**, 238.
3. Olivares, J. A., and Houk, R. S., *Anal. Chem.*, 1985, **57**,
4. Houk, R. S., *Anal. Chem.*, 1986, **58**, 97A.
5. Gray, A. L., *Fresenius z. Anal. Chem.*, 1986, **324**, 561.
6. Date, A. R., and Cheung, Y. Y., *Analyst*, 1987, **112**, 1531.
7. Kawaguchi, H., *Analytical Sci.*, 1988, **4**, 339.
8. Hieftje, G. M., and Vickers, G. H., *Anal. Chim. Acta*, 1989, **216**, 1.
9. Nielsen, F. H., in Draper H. H. Editor, *"Advances in Nutritional Research"*, Plenum Press, New York, 1980, Chapter 6.
10. Combs, G. F., Jr., Levander, O. A., Spaltholz, J. E., Oldfield, J. E., Eds., *"Selenium in Biology and Medicine", Third International Symposium*, van Nostrand Reinhold Co., New York 1987.
11. Lyon, T. D. B., Fell, G. S., Hutton, R. C., and Eaton, A. N., *J. Anal. At. Spectrom.*, 1988, **3**, 265.
12. Robbins, W. B., and Caruso, J. A., *Anal. Chem.*, 1979, **51**, 889 A.
13. Godden, R. G. and Thomerson, D. R., *Analyst*, 1980, **105**, 1137.
14. Date, A. R., and Gray, A. L., *Int. J. Mass Spectrom. Ion Phys.*, 1983, **48**, 357.
15. Powell, M. J., Boomer, D. W. and Me Vicars, R. J., *Anal. Chem.*, 1986, **58**, 2864.
16. Wang, X., Viczian, M., Lasztity, A., and Barnes, R. M., *J. Anal. At. Spectrom.*, 1988, **3**, 821.
17. Janghorbani, M., and Ting, B. T. G., *Anal. Chem.*, 1989, **61**, 701.
18. Ting, B. T. G., Mooers, C. S., and Janghorbani, M., *Analyst*, 1989, **114**, 667.

19. Busheina, I. S., and Headridge, J. B., *Talanta*, 1982, **29**, 519.

20. Ek, P., and Huldén, S. G., *Talanta*, 1987, **34**, 495.

21. Verlinden, M., *Talanta*, 1982, **29**, 101.

22. Franck, A., *Z. Anal. Chem.*, 1976, **279**, 101.

23. Brown, B. M., Jr., Fry, R. C., Moyers, J. L., Northway, S. J., Denton, M. D., and Wilson, G. S.,*Anal. Chem.*, 1981, **53**, 1560.

24. Smith, A. E., *Analyst*, 1975, **100**, 300.

Analysis of Powdered Materials by Laser Sampling ICP-MS

Eric R. Denoyer

THE PERKIN-ELMER CORPORATION, 761 MAIN AVENUE, NORWALK, CT 06859, USA

1 INTRODUCTION

Many solid samples can be analyzed directly by laser sampling ICP-Mass Spectrometry (LS-ICP-MS) without the need for sample preparation. This attribute has generated considerable interest in the laser sampling technique because of the reduction in sample preparation time, sample contamination, and spectral interferences which can be realized. While even loose powders can be sampled directly, stabilizing the powder in a pellet or an adhesive formulation can prove advantageous for laser sampling.

Based on recent investigations in our ICP-MS applications laboratories we have developed several approaches for analyzing powdered materials by LS-ICP-MS. In this paper we wish to share some of our recent experiences in the analysis of soils, ores, sediments, and coal by LS-ICP-MS, and to offer practical observations and recommendations for the optimization and application of the technique.

2 EXPERIMENTAL

For all the measurements described herein, a Perkin-Elmer SCIEX Model 320 Laser Sampler was used in conjunction with a Perkin-Elmer SCIEX ELAN 5000 ICP-MS. A schematic diagram of the LS-ICP-MS equipment used is shown in Figure 1. The instrumentation was operated under the standard operating conditions shown in Table 1.

Pellets of the powdered materials were prepared by mixing three parts powdered sample with one part SpectroBlend™ binder (Chemplex Industries, Inc., Tuckahoe, NY), followed by shaking for five minutes. Approximately 0.5 g of the mixed sample was pressed in a pellet die under 10 tons pressure for one minute.

Table 1 Operating Conditions for ICP-MS Instrumentation

ICP-MS Instrument	Perkin-Elmer SCIEX ELAN 5000
Laser Sampler	Perkin Elmer SCIEX Model 320
RF Power	1100 Watts
Multichannel Dwell Time	20 msec
Integration Time	200 msec
Laser Wavelength	1064 nm
Laser Beam Energy	50 mJ
Pulse Mode	Q-switched
Q-switch Delay	500 μsec
Repetition Rate	Single pulse and 10 Hz

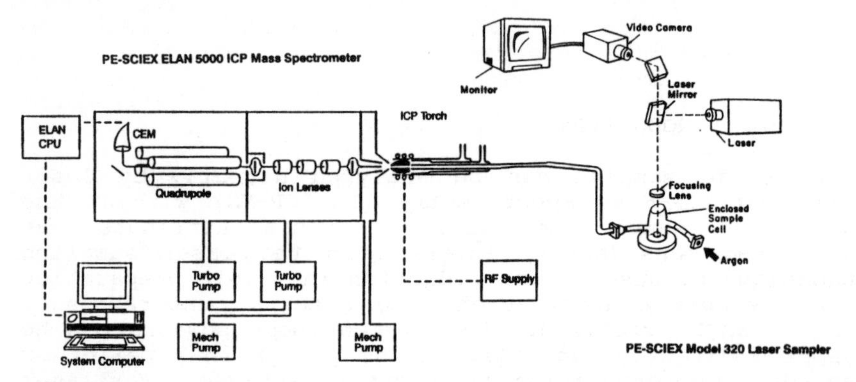

Figure 1 Schematic Diagram of LS-ICP-MS Instrumentation

Plaster pellets were prepared by mixing one part plaster and one part powdered sample with two parts water. This mixture was stirred carefully, left to dry overnight, and weighed. Plaster pellet standards containing known concentrations of analytes were prepared by mixing two parts plaster with one part aqueous 1000 mg/L stock solutions of the desired element(s).

3 RESULTS AND DISCUSSION

There are several possible approaches for analyzing powdered samples by LS-ICP-MS, which include (1) direct analysis of the loose powder, (2) pressing the powdered sample into a pellet either with or without a binder, (3) stabilizing the powder in an adhesive formulation such as plaster or epoxy and (4) preparation of an alkali fusion disc.

Direct Analysis of Loose Powders

In some cases, loose powdered samples can be analyzed directly by LS-ICP-MS. For example, in our laboratories Hager[1] has analyzed powdered zirconia by placing the sample directly into the laser sampler without any prior preparation. Results for semiquantitative analysis of trace elements in the sample were within 30% of the reference value for most elements. Perhaps this approach should not be considered routine because of difficulties with signal stability and with ICP-MS interface overloading which can sometimes be encountered. Nevertheless, it is a rapid, simple approach to the direct analysis of powders, and is especially suitable where a rapid analysis of a limited number of samples is required.

There are several approaches to optimizing the direct analysis of loose powders by LS-ICP-MS. The use of low laser energy tends to minimize the sampling rate and avoid physically disturbing the sample surface. In order to achieve more representative sampling, a larger portion of the specimen can be sampled by defocussing the laser beam and rastering the beam over the sample surface. This is especially important for heterogeneous materials. In fact, the laser focussing lens can be removed altogether to achieve the largest spot size possible.

Sometimes a pre-exposure of the sample is required in order to establish a steady-state signal. This pre-exposure may be longer for powdered samples (e.g. 90-120 sec), and in some cases a steady-state signal may be difficult to achieve. The use of internal standardization can compensate for temporal signal variations and is recommended wherever possible. Another potential source of drift, especially when sampling loose powders, is clogging of the interface cones. Sample loading at the ICP-MS interface can be minimized by minimizing the sampling time, and by inserting a glass wool plug located in the sample transfer line between the sampling cell and the plasma.

Stabilization With Plaster

Powdered samples can be stabilized for LS-ICP-MS analysis by embedding in an adhesive formulation. In this study, the utility of plaster as a stabilizing agent was investigated. Standard pellets were prepared to contain analyte elements at known concentrations. Initial work was performed using the single laser pulse mode producing transient signals as shown in Figure 2. Standard curves were generated from these transient signals (for example see Figure 3). However, with single pulse operation only a very small point on the specimen is sampled. As a result, heterogeneity in both the standard and sample pellets was found to reduce reproducibility.

To minimize this source of variability, the laser was subsequently operated in the continuous pulse mode and was rastered across a 1 cm line on the pellet to obtain a more representative, reproducible signal. In addition to improved sampling, the continuous pulse mode

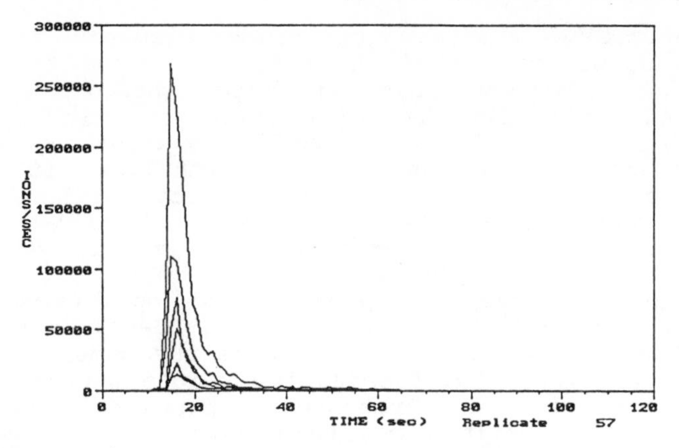

Figure 2 Single laser pulse transient signals of U, Bi, V, Rb, Zn and Ni (order of decreasing intensity) in plaster-stabilized standard pellet

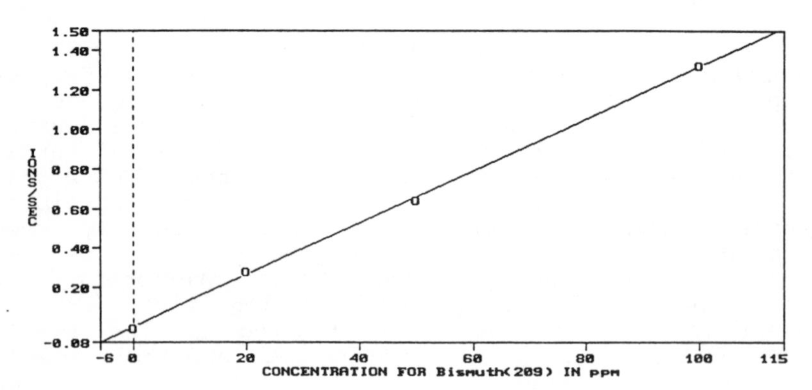

Figure 3 Calibration curve for Bi in Plaster Standard

resulted in higher ion signals compared to the single pulse mode, thereby improving sensitivity and detection limits.

A semiquantitative analysis of the USGS Geological Exploration Reference Material GXR-2 was performed on the plaster-stabilized sample. The TotalQuant spectral interpretation and semiquantitative analysis program was used for the determination. This program, described in detail elsewhere[2], uses the total ICP-MS spectrum to identify spectral features, perform interference corrections, and update the ICP-MS instrument response function. Based on a heuristic, knowledge-based algorithm, TotalQuant then estimates the concentrations of all determinable elements in the sample.

The results of the semiquantitative analysis of the plaster-stabilized GXR-2 are shown in Figure 4. Most of

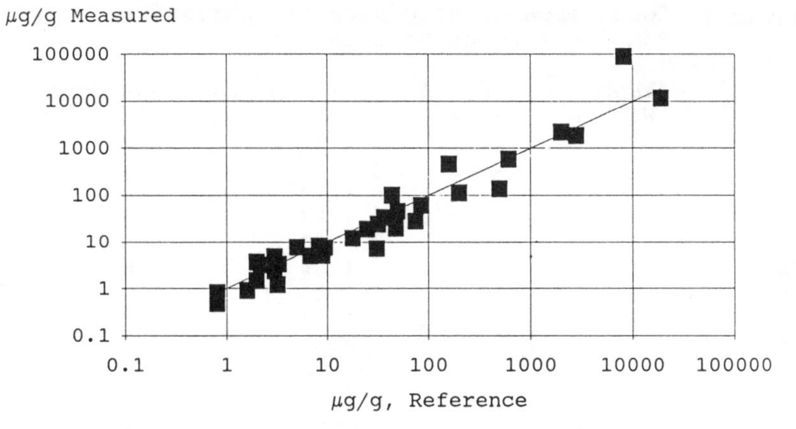

Figure 4 LS-ICP-MS TotalQuant Analysis of Plaster-stabilized GXR-2 Reference Material

the results were generally within 50% of the reference values[3], using a single internal standard element, vanadium, for the whole mass range. However, calcium was in error by a factor of about 10. This error is most likely associated with the very high levels of calcium present in the plaster itself which makes background subtraction difficult. Consequently, plaster stabilization is not recommended for Ca determinations.

Pelletization With a Binder

Paraffin- and cellulose-based binding agents can be used to prepare resilient, somewhat plasticised pellets of many powdered materials. In this study, a commercial binder made of a mixture of paraffin and cellulose was used to prepare pellets of several different powdered sample types.

Reproducibility. The reproducibility of the technique was evaluated by preparing six individual pellets of the USGS GXR-2 reference material. TotalQuant analyses were performed on each pellet, and a comparison was made between the reproducibility of the results determined for repetitive measurements on a single pellet and repetitive measurements on different pellets.

The results for the reproducibility study are shown in Table 2. An F-statistic was used to compare the variance determined for six measurements made within a single pellet and for six measurements made between different pellets. The average variation (% RSD) determined for the six replicate measurements made within a single pellet was 8% for the ten elements studied. This provides an estimate of the measurement precision. The average variation between pellets was 14%, and in all cases except two, the difference between

Table 2 Comparison of precision for pelletization using GXR-2 Reference Material (n=6)

	%RSD Within	%RSD Between	Two-Tailed F-Statistic	95% Conf. Significant?
Be	19	26	2.95	N
Co	4	12	8.29	Y
Rb	5	5	1.13	N
Ag	4	10	7.92	Y
Ba	6	9	2.27	N
Sm	6	12	4.64	N
Dy	11	15	2.05	N
Hf	8	11	1.91	N
Pb	9	14	2.59	N
U	11	23	3.24	N
Ave.	8	14		

the individual % RSD's within and between pellets was not significant (95% confidence). Furthermore, the F-statistics for Co and Ag were only slightly larger than the critical table value. It was therefore concluded that representative pellets could be prepared from the bulk powdered material, and that the pellets were homogeneous within the limits of measurement precision.

Detection Limits. Two approaches were taken for estimating the limits of detection of the technique. Six elements of widely different mass were studied. Sensitivities were determined using the pellets prepared with the USGS GXR-2 reference material. Background signals measured for the argon plasma without the laser firing provided a lower-limit estimate of the blank signals for those elements without significant plasma spectral background structure. This is the case for the six elements studied[4]. Measuring background signals with the laser sampling a pellet prepared from the binding material alone provided an upper limit estimate of the blank signals.

The detection limits estimated from the standard deviations (3-sigma) of these blank measurements are compared in Figure 5. As expected, the values derived from instrument argon background are the lowest, and vary between 0.002 μg/g for U and 0.2 μg/g for Mn. These levels can be achieved in the absence of blank contamination, as is often the case with direct solids analysis. However, if a binding agent is used to prepare a powdered sample for analysis, contamination of the binder can raise the blank signal level, and consequently the detection limit (Figure 5.) Even though the binder was found to contain detectable levels of all the elements measured except Th, the detection limits calculated were still below 1 μg/g for most elements. Nonetheless, is important to select as pure a binding agent as possible for LS-ICP-MS.

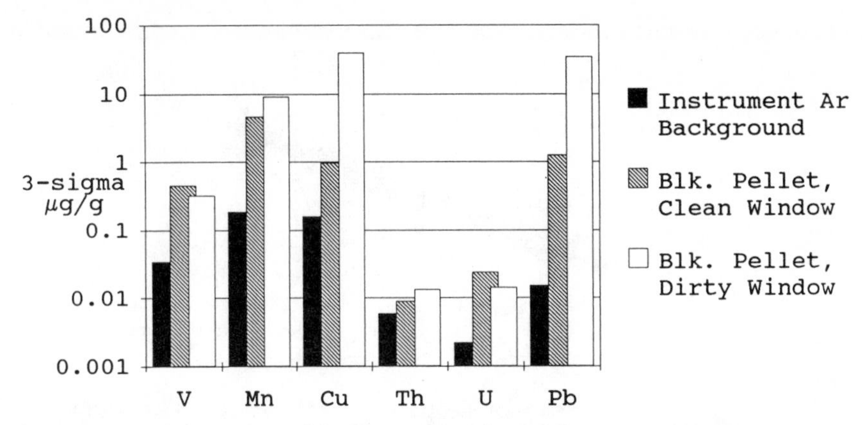

Figure 5 Detection limit comparison for LS-ICP-MS

The other major potential source of background signal is the window of the sample cell through which the laser passes. A worst case situation was simulated by measuring the background signals for the binder pellet using a dirty sample cell window. If the window should get contaminated with material vaporized from samples, subsequent laser pulses can desorb this material, thereby raising the blank signal and the detection limit. Contamination can also degrade accuracy, especially for low-level determinations.

The detection limits calculated using a dirty sample cell window are shown in Figure 5 and are significantly higher for two of the more volatile elements, Cu and Pb, which were desorbed from the window. Detection limit differences between the clean and dirty windows for the other elements were too small to be considered significant. The removable/replaceable sample cell window design used in this work eliminated this possible source of contamination.

Semiquantitative Analysis. The pelletization technique was used to analyze a variety of powdered materials. The TotalQuant program was used for all determinations. For comparison purposes, vanadium was used as the single internal standard in each case. Plots of measured vs. reference values for the USGS GXR reference materials, the NIST SRM 1645 River Sediment, SRM 1632a Bituminous Coal and SRM 1646 Estuarine Sediment are shown in Figures 6-9.

In this study, better analytical accuracy was achieved using pelletization than using plaster stabilization. The average error based on all the detectable elements for the semiquantitative analysis of the GXR series of reference materials (GXR-1,3,4,5, and 6) was less than 40%, except for Ca and Mg which generally fell within a factor of 2-3 of the reference values (see Figure 6.) One result for Rb in GXR-1 fell

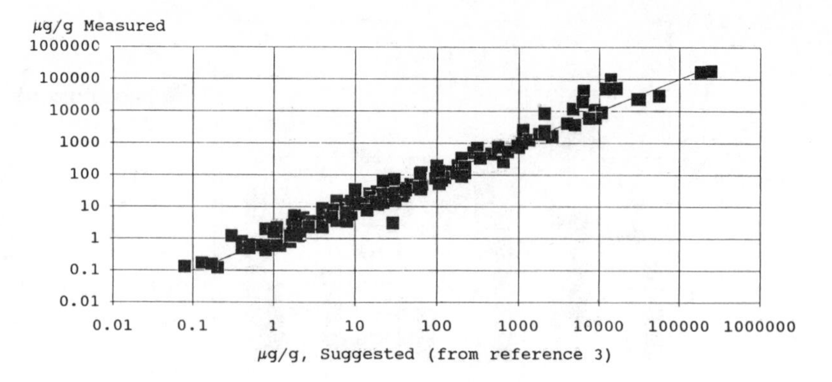

Figure 6 LS-ICP-MS TotalQuant analysis of the USGS GXR series Geological Reference Materials

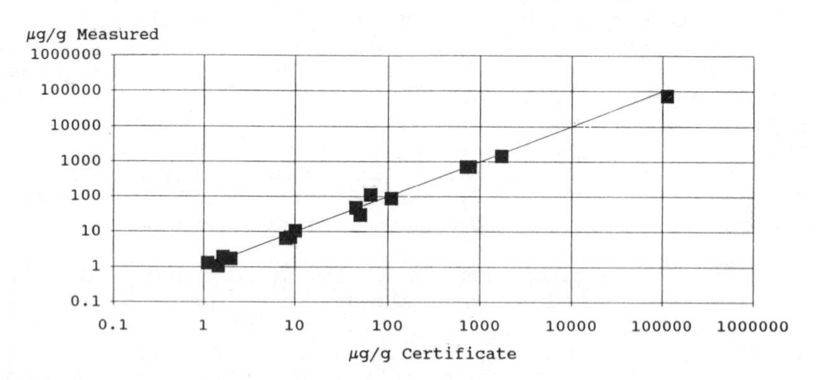

Figure 7 LS-ICP-MS TotalQuant analysis of NIST SRM 1645 River Sediment

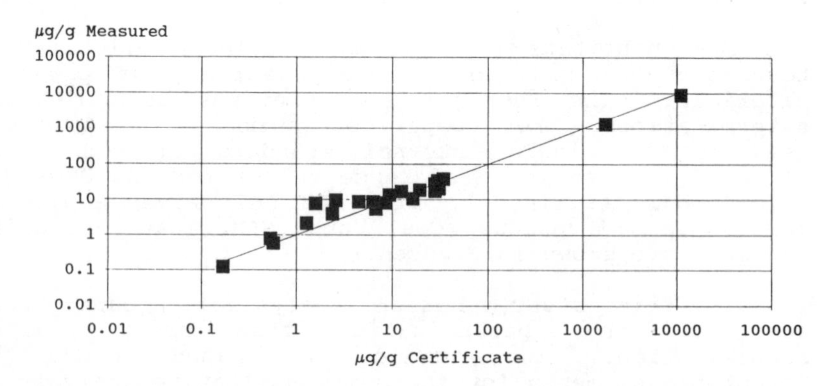

Figure 8 LS-ICP-MS TotalQuant analysis of NIST SRM 1632a Bituminous Coal

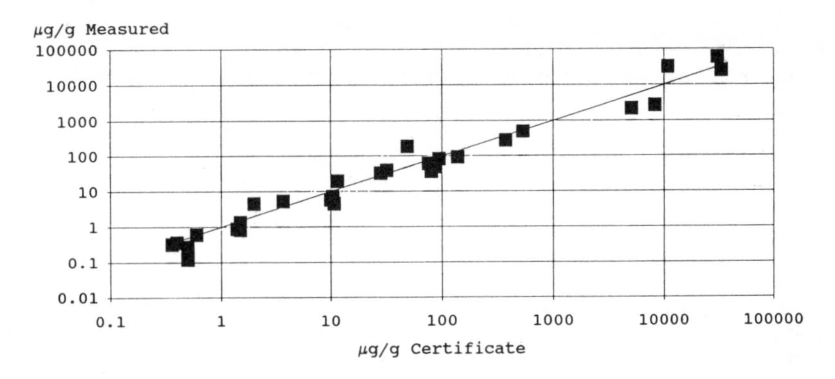

Figure 9 LS-ICP-MS TotalQuant analysis of NIST SRM 1646 Estuarine Sediment

well below the reference value, while all the remaining results for Rb were within the normal accuracy limits. This result cannot be fully explained but may arise from heterogeneity in the GXR-1 sample. The average error obtained for the semiquantitative analysis of the NIST reference materials studied was less than 30% for the certified elements, except for Se in the coal and Ca and Mg in the estuarine sediment. Selenium was within a factor of two and Ca and Mg were within a factor of three of the certified values.

The poorer accuracies for Se, Ca and Mg are thought to result, in part, from the relatively large spectral background correction needed for these elements in ICP-MS. Further improvements are expected using an improved method of blank subtraction for laser sampling under development in our laboratory.

The observed sensitivities (ion counts/sec per µg/g) differed by about a factor of three between the pelletized materials studied. This is not surprising since they differ so widely in matrix composition. However, the use of vanadium (certified in each material) as an internal standard was found to compensate for these sensitivity differences quite effectively for most elements across the periodic table, covering 5-6 orders of magnitude of concentration.

The results for the semiquantitative analysis of these materials are especially encouraging since only a single internal standard was used for each sample. The ability to use only a single internal standard is advantageous for laser sampling because of the potential difficulty in identifying and/or introducing useable internal standards in a solid sample of completely unknown composition. In general, results can be improved by using multiple internal standards, as in solution analysis. Therefore, we are investigating the possibility of adding internal standards during pellet preparation as a means of improving the technique.

4 CONCLUSIONS

Laser sampling ICP-MS can be used effectively to analyze powdered materials. Stabilization of the powdered sample is not strictly necessary but does provide improved reproducibility and control over the measurement. Several approaches can be taken for stabilizing powdered materials, and we find that pelletization is generally faster and easier than fixing in plaster or epoxy for the samples studied.

Internal standardization is recommended wherever possible, but care must be taken in mixing internal standards into the sample since it can be difficult to mix powders homogeneously. Heterogeneity is always a potential problem for direct solids analysis, and grinding the sample to achieve a finely dispersed material of uniform particle size distribution is a good strategy for obtaining a representative analysis. The semiquantitative analysis approach used in these studies (TotalQuant) is a powerful screening tool for rapid sample characterization. Used in conjunction with pelletization and single-element internal standardization, TotalQuant has provided trace element determinations typically accurate to within 30-40% of the known values in the varied sample matrices studied.

REFERENCES

1. J. W. Hager, Seventeenth Annual Meeting of the Federation of Analytical Chemistry and Applied Spectroscopy Societies, Paper No. 294

2. D. Polk, J. Zarycky and R. Ediger, 1990 Pittsburgh Conference and Exposition on Analytical Chemistry and Applied Spectroscopy, Paper No. 85

3. E. S. Gladney, D. R. Perrin, J. W. Owens and D. Knab, Anal. Chem, 51, pp. 1557-1561, 1979

4. S. H. Tan and G. Horlick, Appl. Spectrosc., 40, No. 4, pp.445-460, 1986

Performance Benefits of Optimisation of Laser Ablation Sampling for ICP-MS

I.D. Abell

VG ELEMENTAL LTD, ION PATH, ROAD THREE, WINSFORD, CHESHIRE, CW7 3BX, UK

1. INTRODUCTION

Laser ablation sampling has become an established sampling technique for ICP-MS, primarily because of the advantages in ease, flexibility and speed of analysis which can be achieved for a wide range of sample matrices.

Relatively little work has been reported which explores the optimisation of laser parameters for ablation sampling, even though considerable improvements in performance can be obtained by a brief study of a small number of operating parameters. This paper examines the benefits obtained by simple optimisation procedures which can be completed in several minutes prior to analysis. The effects of laser energy, repetition rate, preablation and laser focus are examined, with respect to the sensitivity and precision of resultant ICP-MS data.

In addition to the examination of bulk analysis performance, the depth profiling resolution of a low energy defocussed beam is determined using samples with thin coatings of known depth.

2. APPARATUS

The instrumentation used was a VG PlasmaQuad Turbo Plus Inductively Coupled Plasma Mass Spectrometer and VG LaserLab Plus laser ablation sampling accessory. The laser beam was generated by a Nd:YAG pulsed laser operating in the infrared region at 1064 nm. The beam was focused onto the sample surface through the window of a quartz sample cell, flushed by approximately 1 l/minute of argon carrier gas. Ablated aerosol was carried through nylon and tygon tubing to the injector inlet of the ICP torch. Viewing of samples was accomplished by high resolution CCD video camera and monochrome monitor.

Time Resolved Analysis (TRA)

TRA is a method of scanning or peak jumping data acquisition which places full mass range data into sequential time slices of user-definable width. TRA software was used here to give a rapid graphic assessment of the effect of variation of individual parameters of laser operation. As a parameter was changed at regular time intervals, its effect was logged by the TRA acquisition. TRA was also used in depth profiling using laser sampling at a fixed energy per shot and fixed repetition rate. In this case, the 'time' axis of the TRA corresponded to the depth of penetration of the laser into the sample, creating a pseudo - Dimensionally Resolved Analysis (DRA).

Laser Mode

All data relates to the operation of the laser in Q-switched mode, pulse length 15ns.

Optimisation of Laser Energy and Repetition Rate

At a repetition rate of 4Hz, the energy per shot of laser sampling was gradually increased during ablation of a nickel alloy BCS 346a. TRA data of time slice width 3 seconds was acquired throughout. There was an increase in sensitivity of analysis of all trace elements with increasing laser energy, as shown for the case of aluminium in Figure 1. It is evident, at higher energies, signal stability was worsened, indicated by the more irregular intensities of sequential time slices. Although not fully quantified, the graphics of TRA have quickly shown that above 0.2 J per shot, greater sensitivity is however only achieved with the loss of signal stability.

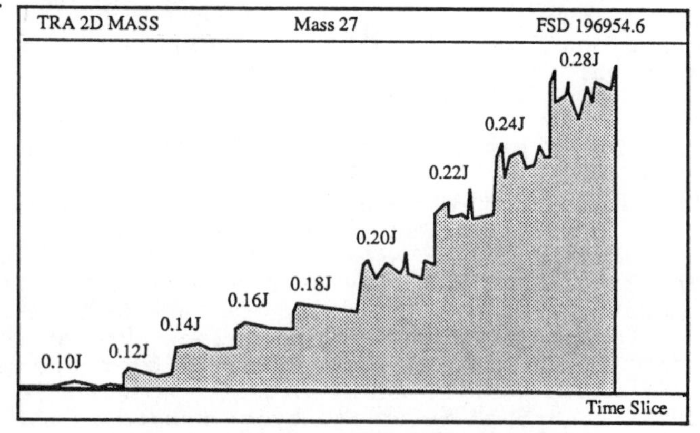

Figure 1 Variation in signal with Increasing Laser Energy - Aluminium in Low Alloy Steel.

Having now chosen a compromise optimum energy per shot of 0.18J, the effect of increased repetition rate was then investigated using TRA acquisition and the sample BCS 346a. Operating at 0.18 J per shot, the laser repetition rate was increased in increments of 2Hz, between 2Hz and 10Hz. Each repetition rate was maintained for a period of two minutes, giving a total experimental time of ten minutes. A number of trace element isotopes were monitored by TRA, yielding a clear graphical illustration of the effect of repetition rate variation across the mass range. The TRA plot shown in Figure 2 clearly shows that sensitivity of antimony analysis increased in direct proportion to the laser repetition rate, and that at this particular energy per shot, there is no significant stability loss up to a 10Hz repetition rate.

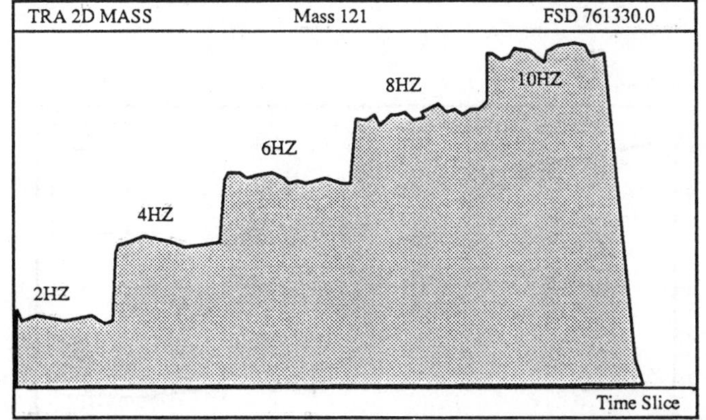

Figure 2 Variation in Signal with Laser Repetition Rate - Antimony in Low Alloy Steel

TRA has been used here to find a quick compromise optimum of energy and repetition rate, with respect to both the sensitivity and stability of the analytical data. The total time for this optimisation was of the order of fifteen minutes.

Optimisation of Laser Focus

For this exercise, conventionally integrated data was used to establish the effect of laser focusing on precision as well as its effect on sensitivity.

For a steel and a nickel alloy matrix, the effects of laser focus on both sensitivity and short term analytical precision were examined. For nickel alloy BCS 346a, Figures 3 and 4 clearly show that optimal behaviour for both sensitivity and precision occur at the same defocus setting. The units of defocus are the extent to which the laser was defocused below the surface of the sample, as shown in Figure 5. In Q-switched mode, the focused laser beam generates a brief laser induced plasma of ionised argon, sample vapour, and free electrons, immediately

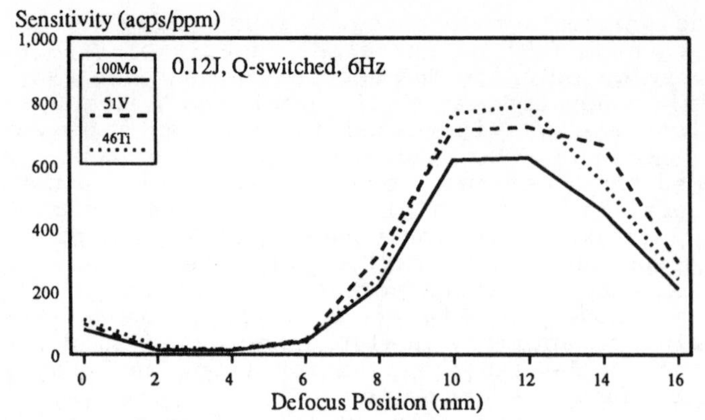

Figure 3 Focus Optimisation - Laser Sampling of BAS 346a

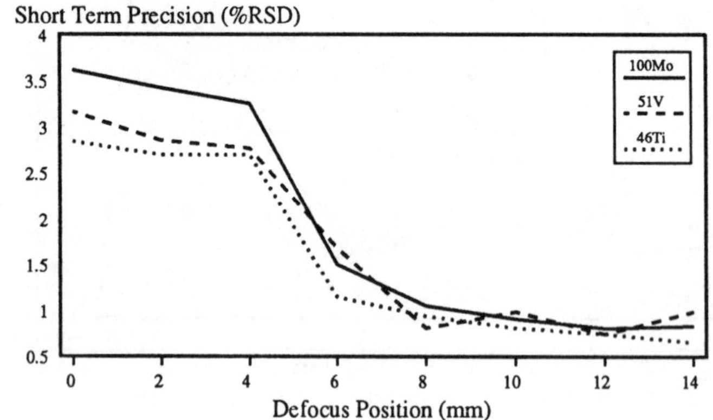

Figure 4 Short Term Precision/Laser Focus - BAS 346a Nickel Alloy

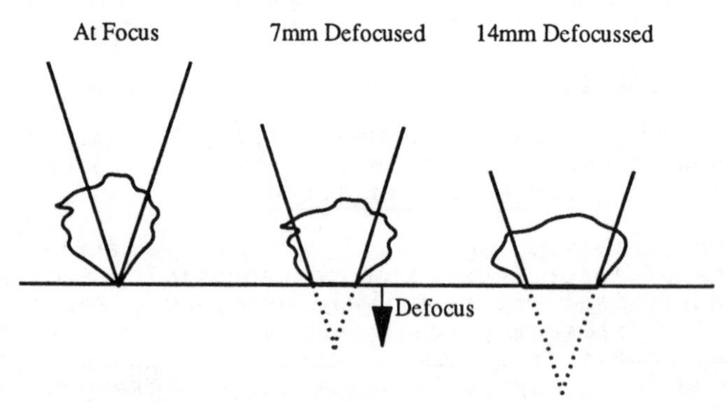

Figure 5 Laser Induced Plasma Formation Effect of Focus

above the sample. Much of the energy of the Q-switched beam reaches the sample surface indirectly via the formation of the laser induced plasma. The degree of focus of the beam influences the way in which the plasma propagates and has a very significant effect upon the efficiency of coupling of energy to the sample.

For a wide range of matrices, the optimal precision and sensitivity of analysis were to be found at the same defocus condition. It is reasonable to expect therefore, that monitoring sensitivity only, again perhaps using TRA methods, would lead also to a precision optimum, without actually needing to measure precision. Differing matrices will have different optimum defocus conditions. Table 1 lists a range of defocus optima found for a range of matrices. The table lists both sensitivity and precision optima. The variation in optimum conditions between materials is obviously wide, and so the determination of an optimum for a previously unknown matrix is likely to yield significant performance benefits.

Table 1 Typical Optimum Defocus Conditions for Sensitivity and Precision

Matrix	Defocus Optimum (Sensitivity)	Defocus Optimum (Precision)
NBS Glasses	4mm	6mm
Low Alloy Steel	12mm	12mm
Ni/Fe/Cr Alloys	14mm	12mm
High Purity Copper	3mm	4mm
Silicon	10mm	9mm
P.E.T. (Polymer)	7mm	6mm
Quartz (Suprasil)	0mm	1mm

Optimisation of Sample Preablation

Analytical precision has been found to be significantly improved by preablating the sample surface, using the laser under the same conditions which will be used for analysis.

Typical preablation times are of the order of 3 to 4 minutes at medium laser energies and repetition rates. Figure 6 shows the improvement in short term precision achieved by preablation of a zirconium alloy.

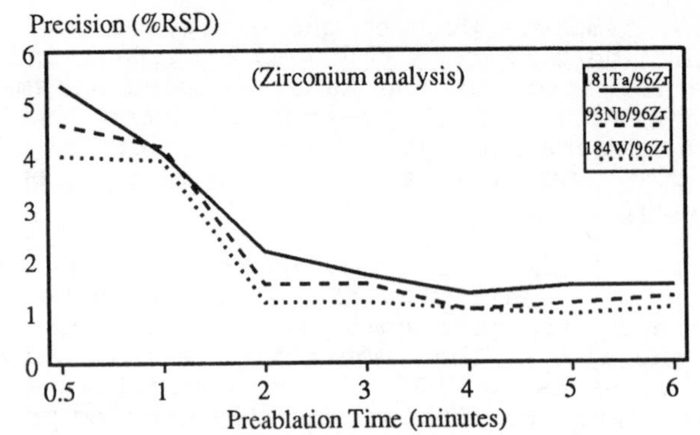

Figure 6 Effect of Sample Preablation on Precision of Measurement

An illustration of the benefit of a brief period of optimisation is shown in Table 2. This table lists the medium and long term precisions of a number of minor element determinations in a steel sample, ratioed to the ^{57}Fe isotope.

The laser energy, repetition rate, defocus condition and preablation period had all been optimised prior to acquisition of this data.

Table 2 One Hour and Four Hour Precisions.
$M^+/^{57}Fe$ - NBS 1765
(4 minutes preablation prior to .on ofsis)

Element	Content (%)	1 Hour Precision ($M/^{57}Fe$)	4 Hour Precision ($M/^{57}Fe$)
V	0.004	2.7	1.9
Cr	0.05	1.6	1.0
Mn	0.14	1.4	1.4
Co	0.002	1.3	2.1
Ni	0.15	1.9	2.2

(ALL VALUES % RSD)

Depth Profiling Resolution

Low energy, low repetition rate laser sampling was used to ablate material from three coated steel samples. The coating materials and thicknesses were known and so the relative rates of penetration through the materials were quantifiable. During sampling, TRA data was acquired, with a time slice width of 4 seconds. Examples of the TRA graphics display are shown in Figure 7.

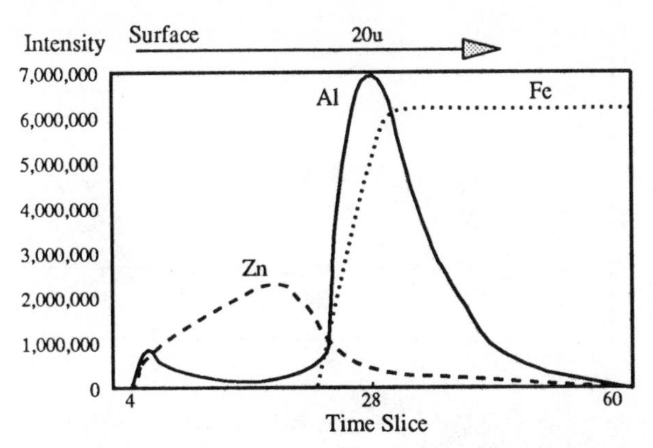

Figure 7 Depth Profile of Zinc Coating on Steel Substrate

The zinc layer on this sample was 20 microns thick, so it can be concluded that the rate of penetration of the laser through the layer was 0.15 microns per shot. The other two samples were 10 micron layers. One was also zinc and the other was aluminium. The penetration rates for these layers were 0.13 microns per shot (Al) and 0.14 microns per shot (Zn). The TRA data shows also that there is a thin layer of approximately 5 microns of aluminium between the steel substrate and the zinc coating.

For one of the zinc coatings, TRA data for the coating was integrated and used to calculate the composition of the coating material. The calculated and nominal compositions are compared in table 3. The crater produced during depth profiling of this sample is shown in figures 10 and 11. There is a distinct boundary between the ablated region and the unaffected surface, and the crater surface is relatively flat, which enhances the ability of the laser to reliably resolve specific layers within the coating.

Table 3 Determination of Composition of 10 micron coating by LA ICP-MS

Element	Nominal	Found
Zn	90	84
Al	0.13	0.21
Fe	10	15.7
Sb	0.06	0.03
Si	TRACE	0.078

(All values in weight %)

Figure 8 Depth Profiling Crater - Defocused Q-Switched Ablation, 0.07J per Shot

3. CONCLUSIONS

1. Using Time Resolved Acquisition (TRA), with time slice widths of up to 3 seconds, the energy and repetition rate of laser ablation sampling can both be optimised with respect to the sensitivity of analysis and the stability of aerosol generation. This optimum set of conditions may be determined for a previously unknown matrix in less than fifteen minutes using only a visual inspection of the TRA graphics.

2. Defocusing the beam for Q-Switched laser sampling can produced up to an order of magnitude increase in sensitivity, depending on the matrix being sampled. The optimum with respect to sensitivity is often very close to the precision optimum defocus value, and so both can be attained by measuring only the sensitivity / defocus relationship.

3. Preablation of sample prior to data acquisition can significantly improve the short term and long term reproducibility of analysis. No particular enhancement in sensitivity is generally found after sample preablation.

4. Medium and long term precisions of analysis of the order of 2% to 3% RSD may be expected from some matrices following brief optimisation of parameters. Preablation prior to acquisition for each sample will extend total analysis time to typically five or six minutes to achieve this precision.

5. Depth profiling using defocussed, low power Q-switched ablation can generate a depth resolution of less than 0.2 microns per shot, yielding detailed information as to the spatial distribution of trace, minor and major elements. For zinc and aluminium coatings, the rate of penetration for different coating thicknesses is reproduced, so time resolved analysis may be used to record full mass range data to give a reliable graphical indication of each individual element distribution.

6. Time resolved analysis data may be integrated for specific regions of sample in order to generate quantitative or semiquantitative analysis of specific areas or layers.

Subject Index